INORGANIC CHEMISTRY

INORGANIC CHEMISTRY

Editor
RUBY GARG

KALPAZ PUBLICATIONS
DELHI-110052

INORGANIC CHEMISTRY

(Chemistry)

ISBN: 81-7835-050-5

Published in 2001 in India by
Kalpaz Publications
C-30, Satyawati Nagar
Ashok Vihar Phase-III
Delhi-110052
E-Mail: kalpaz@hotmail.com

Laser Typesetting at : A.S. Printers
Printed at : Mehra Offset Printers, New Delhi

Preface

The term 'inorganic chemistry' originally meant to non-living chemistry, and it was that part of chemistry which had arisen from the arts and recipes dealing with minerals and ores. It began by finding naturally occurring substances that had useful properties, such as flint or chert that could be worked into tools. This search continues, but now it is included in the science of mineralogy and geology. Inorganic chemistry deals more with the changes that can be effected in materials.

Inorganic chemistry interacts with organic, physical and even biological chemistry. It is impossible in a single book to do more than scratch the surface of inorganic chemistry: structure, reactivity, catalysis, thermodynamic stability, symmetry, experimental techniques, gas phase, solution and solid-state chemistry; they are all part of the process. Each topic of this book attempt to provide the reader with sufficient basic knowledge of the structure and reactivity of inorganic system to ensure a more comprehensive understanding.

This book provides an authoritative account of every aspect of inorganic chemistry of current interest and demonstrates progress in this subject that has been made in the recent past. It gives emphasise on the primary facts of inorganic chemistry, and have organised the facts of chemical structure and reactivity in a way that emphasises the descriptive approach to the subject. The objective of this book is to make the concepts and methods of inorganic chemistry clear and interesting to students. A complete insight into the subjects becomes possible only when the student understands the basic principles. The chapters have been arranged in a conventional way, as it may provide good for students to pass from one to another with good continuity. The order of the techniques included is conventional and would be liked by students.

Editor

Contents

Chapter 1

Properties of Inorganic Compounds

Small quantities of oxygen were introduced and sparking continued until the volume reached a minimum when excess oxygen was absorbed by a solution of potassium polysulphides (liver of sulphur). A small bubble of gas remained which was neither nitrogen nor oxygen and to quote Cavendish, 'we may safely conclude that it is not more than 1/120 part of the whole.' Cavendish had isolated a mixture of inert gases. Here, the matter rested for over 100 years.

The names and some physical characteristics of the inert oases and the composition of the Earth's atmosphere, which contains all of the inert gases except radon, is given in Table 1. Radon is an isotopic mixture of radon.

Randon is an isotopic mixture of randon, thoron and actinon and was once called emanation (1900) because these gases are respectively emanations of radium, thorium, and actinium, i.e. produced during the radioactive decay of these elements. All are radioactive. The half-life of thoron is 54.5 s of actinon, 3.9 s, and of radon, 3.8 days. It has been estimated that c. 60% of lung cancer deaths are caused by radon, the single most important source of radiation exposure, produced by radioactive decay in the granite of SW England and NE

Scotland. The National Radiological Protection Board recommend that the annual dose of radiation received at work should not exceed 15 mSv. It is estimated that ten times as many people experience more radiation at home every year than do people working in industry.

Table 1

Composition of the Earth,s atmosphere/parts in 105 of dry air

By volume	Gas	By mass
780800.00	Nitrogen	755200.00
209500.00	Oxygen	231500.00
9320.00	Argon	128000.00
300.00	Carbon dioxide	460.00
18.00	Neon	12.00
5.20	Helium	0.72
1.50	Methane	0.80
1.10	Krypton	3.00
0.50	Dinitrogen oxide	0.80
0.50	Hydrogen	0.04
0.40	Ozone	0.70
0.09	Xenon	0.40

The density of nitrogen prepared from air by removal of all other known constituents, and of nitrogen prepared from chemical sources: the respective values were 1.2572 g dm^{-3} and 1.2505 g dm^{-3}, under standard conditions, 273 K and 101 kPa. The density of atmospheric 'nitrogen' was the higher by about 1 part in 200 which was more than the degree of experimental error.

The discrepancy was due to an unknown impurity in atmospheric nitrogen, also started investigating the problem. He started with dry air, from which carbon dioxide had been removed, and removed oxygen by repeated passage over heated copper, which was converted to the oxide. He passed the remaining gas over heated magnesium, which combined with the nitrogen to form magnesium nitride. This was continued until the residual gas had a constant relative density.

The gas had a spectrum which was new and quite unlike that of nitrogen. It was chemically inert. Lord Rayleigh had continued his experiments of sparking air with oxygen and had obtained small quantities of the gas. They named the gas argon and published their results in a joint paper. Ramsay looked for other sources of argon and his attention was drawn to a mineral cleveite from which Hildebrand had released what seemed to be nitrogen.

Having removed the nitrogen, Ramsay found that samples of the gas were composed mostly of a gas with a spectrum different from that of argon or nitrogen. This was identified with spectrum lines detected by Janssen in the solar spectrum during an eclipse in 1868 and interpreted by Lockyer as being due to an element unknown on the Earth, which he had named helium (Greek: *helios* = sun), the suffix *-ium* indicating that a metallic nature was assumed. Helium is associated with many radioactive minerals and may be released by heating them. It is formed by radioactive decay and a helium nucleus is identical with an a-particle. With the appearance of a group 0 (zero valency) in the Periodic Table to accommodate new elements with unusual characteristics, Ramsay realized there was probably an undiscovered element of similar properties with an intermediate atomic mass and probably others which had larger atomic masses. At about this time, liquid air had been made in bulk by.

Hampson in England and Linde in Germany. By controlled evaporation of liquid air, various components could be obtained. From the residual few millilitres of liquid left when nearly all of about one and a half litres of liquid air had been allowed to evaporate, they isolated krypton, identifying it by its distinct spectrum. Afterwards, neon and xenon were also isolated and identified spectroscopically. In 1907, up to 1.84% by volume of helium was discovered in natural gas by Cady and McFarland in Kansas, USA.

Although up to 7-8% by volume has been found, quantities nearer 1074 are more usual. The natural gas is associated with the petroleum regions of the USA and consists mostly of methane, the helium having (presumably) collected as a result of radioactivity. Except for radon, the other inert gases are produced from the atmosphere. Because of their inert nature, the gases of this family have interesting and specialized uses. Deep-sea divers require the supply of oxygen under considerable pressure. If air is supplied at high pressure, large quantities of nitrogen are dissolved in the blood stream and the sudden release of this by bringing the diver up rapidly would result in blockage of the circulation by bubbles of nitrogen.

A mixture of helium, which is much less soluble, and oxygen is used instead of air and the diver may be raised quickly. Helium (b.p. -268.9°C) is used in low-temperature studies and in providing an inert atmosphere for certain welding processes. Discharge tubes filled with neon are used for illumination and for advertisement signs, the red-orange colour being modified by additions of other rare gases, mercury, or the use of coloured glass as desired. With 15070 nitrogen, argon is used in gas-filled electric lamps and in fluorescent lighting tubes. It is also used as an inert atmosphere for welding. Radon has been used in the treatment of cancer because of its radioactivity.

Classification of the elements in the 20th century

An atom is believed to be composed of a relatively mall,

dense, positively charged nucleus, which is' formed from protons and neutrons, surrounded at a distance by sufficient negatively charged electrons to leave the whole atom electrically neutral. In theory, elements may be formed by constructing atoms of the appropriate atomic number by using this number of protons in the nucleus with the required number of neutrons as a sort of nuclear ballast and surrounding the nucleus at a distance with enough electrons to preserve electrical neutrality,. the number of electrons being equal to the atomic number. Nothing has been said about the configuration adopted by these electrons.

Many physical properties depend to some degree on the size of the atoms of the element under review. The size or volume of an atom will be determined by the distance at which the outermost electrons lie from the nucleus. The atomic volume is a convenient but tough way of measuring an atom's size. It equals the volume occupied by the molar mass of the element and depends on the geometrical packing of atoms in the solid, especially where polyatomic entities occur, and on the conditions under which the density of the solid is determined.

When plotted against atomic mass, the atomic volumes of successive elements form a periodic curve. The fundamental quantity for an atom is now known to be the nuclear charge, proton number or atomic number, not the atomic mass. Although atomic volume, as defined and measured here, depends on so many factors, it seems reasonable to suppose that this periodicity is related to the electron distribution in the atoms selected.

When the light from a sodium vapour lamp is examined in a spectroscope a bright yellow line is observed. On narrowing the slit by which light enters the spectroscope, two lines close together may be observed.

Periodic Table of the elements

I A (1)	II A (2)	III A (3)	IV A (4)	V A (5)	VI A (6)	VII A (7)	VIII A (8)	VIII A (9)
1 H 1.00797 Hydrogen								
3 Li 6.941[b,c] Lithium	4 Be 9.01218 Beryllium							
11 Na 22.98977 Sodium	12 Mg 24.305 Magnesium							
19 K 39.0983 Potassium	20 Ca 40.08 Calcium	21 Sc 44.9559 Scandium	22 Ti 47.90 Titanium	25 V 50.941 Vanadium	24 Cr 51.996 Chromium	25 Mn 54.9380 Manganese	26 Fe 55.847 Iron	27 Co 58.9332 Cobalt
37 Rb 85.467 Rubidium	38 Sr 87.62 Strontium	39 Y 88.9059 Yttrium	40 Zr 91.22 Zirconium	41 Nb 92.9064 Nobium	42 Mo 95.94 Molybdenum	43 Tc 98.9062[d] Technetium	44 Ru 101.07 Ruthenium	45 Rh 102.9055 Rhodium
35 Cs 132.9054 Cesium	56 Ba 137.34 Barium	57 La 138.905 Lanthanum	72 Hf 178.49 Hafnium	73 Ta 180.948 Tantalum	74 W 183.85 Tungsten	75 Re 186.2 Rhenium	76 Os 190.2 Osmium	77 Ir 192.2 Iridium
87 Fr Francium	88 Ac 226.0254[d] Radium	89 Ac 227.0278 Actinium	104	105	106	107	108	109

58 Ce 140.12 Cerium	59 Pr 140.907 Praseodymium	60 Nd 144.24 Neodymium	61 Pm Promethium	62 Sm 150.35 Samarium
90 Th 232.0381 Thorium	91 Pa 231.0359 Protactinium	92 U 238.029[c] Uranium	93 Np 237.0482 Neptunium	94 Pu Plutonium

VIII A (10)	I B (11)	III B (12)	III B (13)	IV B (14)	V B (15)	VI B (16)	VII B (17)	VIII B (18)
								2 He 4.00260 Helium
			5 B 10.811[b] Boron	6 C 12.01115 Carbon	7 N 14.0067 Nitrogen	8 O 15.9994[b] Titanium	9 F 18.9984 Fluorine	10 Ne 20.179 Neon
			13 Al 26.98154[b] Aluminum	14 Si 28.086 Silicon	15 P 30.97376 Phosphorus	16 S 32.064[b] Sulfur	17 CI 35.443 Chlorine	18 Ar 39.948 Argon
28 Ni 58.69 Nickel	29 Cu 63.546[b] Copper	50 Zn 65.377 Zinc	31 Ga 69.72 Gallium	32 Ge 72.59 Germanium	33 As 74.9216 Arsenic	34 Se 78.96 Selenium	35 Br 79.904 Bromine	36 Kr 83.80 Krypton
46 Pd 106.4 Palladium	47 Ag 107.868 Silver	48 Cd 112.40 Cadmium	49 In 114.82 Indium	50 Sn 118.69 Tin	51 Sb 121.75 Antimony	52 Te 127.60 Tellurium	53 I 126.9045 Iodine	54 Xe 131.30 Xenon
78 Pt 195.09 Platinum	79 Au 196.9665 Gold	80 Hg 200.59 Mercury	81 TI 204.38 Thallium	82 Pb 207.19[b] Lead	83 Bi 208.9804 Bismuth	84 Po Pollonium	85 At Astatine	86 Rn Radon

63 Eu 151.96 Europium	64 Gd 157.25 Gadolinium	65 Tb 158.9254 Terbium	66 Dy 162.50 Dysprosium	67 Ho 164.9304 Holmium	68 Er 167.26 Erbium	69 Tm 168.9342 Thulium	70 Yb 173.04 Ytterbium	71 Lu 174.97 Lutetium
95 Am Americium	96 Cm Curium	97 Bk Berkelium	98 Cf Californium	99 Es Einsteinium	100 Fm Fermium	101 Md Mendelevium	102 No Nobelium	103 Lw Lawrencium

This spectrum, produced by excitation of sodium atoms, is called a line spectrum or an atomic spectrum. Some are very much more complex. Atomic spectra may be produced by electrical discharge, arcing or by heating and are emitted by atoms in the vapour state. On the other hand, light from an incandescent solid, on spectroscopic analysis, forms a continuous spectrum. Some metals may be identified by the characteristic colours, or, if necessary, their spectral lines, given under suitable conditions in qualitative analysis.

Chlorides of the metals are used. They are usually relatively volatile and easily obtained by mixing a powdered solid compound with concentrated hydrochloric acid. A platinum wire, moistened with the mixture and held in a (non-luminous) Bunsen flame may impart a characteristic coloration to the flame. Molecules give a different type of spectrum, in which successive bands appear, called a band spectrum. On resolution, each band appears as a number of close-packed lines. Spectra may appear in the infrared, ultra-violet and visible parts of the spectrum.

At present, we confine ourselves to the knowledge that suitably excited atoms give line spectra. Figure 1 shows line spectra for lithium, sodium and potassium. Wave-lengths are expressed in nanometres (rim = 10-9 m) an d picometres (pm = 10^{-12} m). The emission of lines of characteristic wave-lengths must be associated with definite energy changes. Further, it is found that elements from the same family often have similar atomic spectra. The atomic spectrum for hydrogen has proved to be relatively the most simple and the wave-lengths (A) of the multiplicity of lines in ultra-violet, visible and infrared regions are found to obey the simple formula:

$$1/\lambda = R_H\left(\frac{1}{n^2_1} - \frac{1}{n^2_2}\right)$$

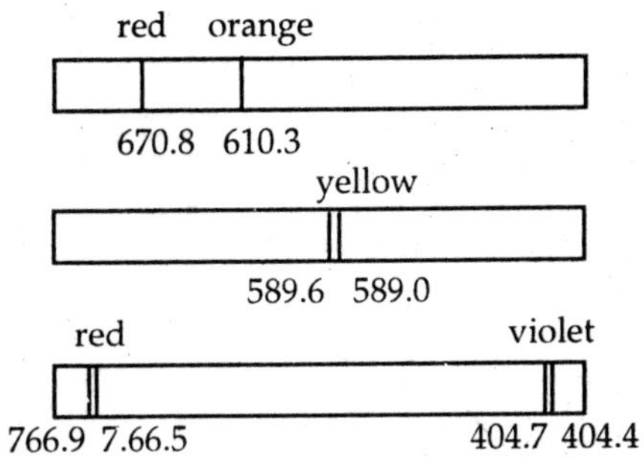

Fig. 1. Line spectra for lithium, sodium and potassium showing colours and wave-length/nm

where n_1 and n_2 are integral and n_1 assumes a fixed value for all lines of a given series. Table 2 gives the values of n_1 and n_2 for the different series, each of which is named after the discoverer, together with the region to which the series belongs. RH is the Rydberg constant for hydrogen-observed value = 10 967 818 m^{-1}. The continuous spectrum of an incandescent solid is in no way dependent on the nature of the solid.

The distribution of energy or intensity of radiation is characteristic of the temperature. For a perfect radiator, called a black body, the total energy emitted per second is proportional to the fourth power of the absolute temperature (Stefan's Law) and the maximum value of the radiation for unit wave-length interval is at a wave-length which is inversely proportional to the absolute temperature (Wien's Law). In order to explain these observations the Quantum Theory was advanced by Planck in 1900. Planck suggested that energy is emitted or absorbed in discrete units or quanta: each quantum is a quantity of energy (E) which is proportional to the frequency v (greek letter, nu) of the radiation and is given by the formula:

$$E = hv$$

Table 2

The atomic spectrum of hydrogen

Series	Years of discovery	region	n_1	n_2
Lyman	1906	Ultra-violet	1	2, 3, 4,...
Balmer	1885	Visible	2	3, 4, 5,...
Paschen	1908	Infra-red	3	4, 5, 6,...
Brackett	1922	Infra-red	4	5, 6, 7,...
Pfund	1924	Infra-red	5	6

where h = 6.6256 X 10^{-34} Js (Planck's constant). This hypothesis has developed into a theory which is one of the most important in physics and chemistry. Thus, quanta (*hv*) emitted as infra-red radiation from a hot body will be smaller than those emitted as the body reaches incandescence. Einstein extended these ideas to the transmission of energy by atoms and suggested that radiation travelled in quanta.

Each quantum is called a photon and represents a small increment of energy ($E = hv$) travelling with the velocity of light ($c = 3 \times 10^8$ in s^{-1}). The Rutherford atom had been pictured as a miniature solar system upon which was imposed a distribution of electricity resulting in a positively charged nucleus exerting a force of attraction on the negatively charged electrons revolving in their orbits. Rapid motion was the best explanation afforded for the apparent stability of electrons remaining in orbit against the pull of attraction from the heavier nucleus. This brought in another difficulty since such an arrangement would require, from a classical theory, continuous radiation with consequent spiralling of the electrons into the nucleus as energy was lost.

However, novel suggestions, which have resulted in the Quantum Theory being extended to atomic structure, were put forward in 1913 by Bohr. It was postulated that electrons may occupy orbits (at a distance from the nucleus) from which no energy is absorbed or emitted. These were called stationary states and are now referred to as energy levels. A further postulate required that absorption of a suitable quantum of energy results in the virtually *instuntaneous* promotion of an electron from an inner to an outer orbit, from the normal or ground state of least energy to a higher level.

The energy absorbed is related to the frequency of radiation absorbed by $\Delta E = hv$. When an electron returns from this energy level to the original level, the same quantity of energy is emitted and appears as radiation of the same frequency as before ($\Delta E = hv$). By having a number of energy levels into which an electron can be promoted or fall without necessarily returning directly to its original level, Bohr developed a theory of the hydrogen atom which explains the series of lines observed in the atomic spectrum of the gas.

Figure 2 shows a pictorial representation of the formation of the various series making up the full atomic spectrum of hydrogen. Electrons are found in quantum shells or electron shells, associated with energy levels. Each energy level is numbered with a principal quantum number. Alternatively, the letters, used originally to describe X-ray spectra, may be used instead: K, L, M, ... Other experiments have confirmed the concept of energy levels.

The position with other atoms is much more complex and is treated in the subject of wave mechanics or quantum mechanics. The treatment regards the atom as a mathematical concept without regard for a simple pictorial representation because it is now believed that the electron cannot be regarded as a conventional material particle.

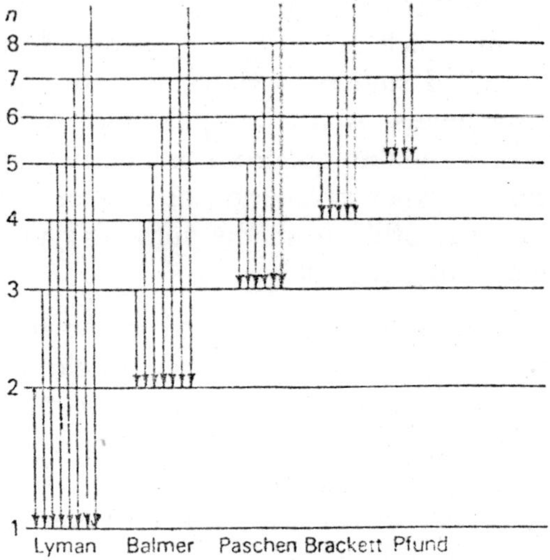

Fig. 2. The formation of the various series of lines belonging to the atomic spectrum of hydrogen

The measurement of ionization energies, providing firm evidence for the orderly expansion of the electronic configuration of atoms and necessary associated theory is described later in this chapter after the Periodic Table has been established according to the principles set out so far. Events in the gas discharge tube were interpreted in terms of a particle concept (=the electron) while later developments in theory and experiment have resulted in the need to explain some further phenomena in terms of a wave character. Just as with light, the electron is the name given to a group of linked phenomena which are best explained sometimes in particle terms and in other instances in wave terms. Now, it is quite meaningless to ascribe physical properties to things unless measurements can be made.

The electron is so small that any attempt to measure its speed, or the related quantities, momentum or energy (by using suitable radiation perhaps) would result in the disturbance of the electron (because of collision with at least one photon of radiation, photon being the name given to one quantum of light) and uncertainty about its position. This is an aspect of the Heisenberg Uncertainty Principle; it is unconnected with ordinary experimental errors in any way, and is implicit in the study of sub atomic phenomena. However, in chemistry, we may safely visualize the electron as a particle if we do not insist on defining exactly its speed and position in orbit.

The mathematical treatment involves wave equations and the wave function may best be interpreted as giving the probability of finding an electron at a particular position. While we find it useful to regard electrons as tiny particles, the overall picture of an atom, when built from the main sub-atomic units, should be of a positively charged nucleus surrounded by a diffuse cloud of negative electric charge ('electron-c loud'), gradually thinning out to the periphery of the atom. The probability of finding an electron is greatest where the cloud is densest. Having dwelt on the inadequacy of physical models, we now turn to the question of configurations adopted by the electrons in atoms.

Soon after the discovery of the electron the link between chemical properties and electronic configuration was apparent and the idea grew that the lack of chemical reactivity shown by inert gases was due to completed shells of electrons. Starting here, a scheme was developed to explain the position of elements in the Periodic Table in terms of electronic configurations. Associated with this theory are Langmuir, Bury, Bohr and Pauli. The electronic configurations of the inert gases are listed in Table 3. Helium contains two electrons, when in its ground state, occupying the first or K electron shell which cannot expand further. Neon has eight electrons in the second or L electron shell besides the two in the K shell. Argon has a further eight electrons in the third or M shell. This straightforward expansion is now modified.

Table 3

Electronic configurations of the noble gases

Name	Symbol	Atomic number	Designation of electron shell 1 K	2 L	3 M	4 N	5 O	6 P
Helium	He	2	2					
Neon	Ne	10	2	8				
Argon	Ar	18	2	8	8			
Krypton	Kr	36	2	8	18	8		
Xenon	Xe	54	2	8	18	18	8	
Radon	Rn	86	2	8	18	32	18	8

In passing from argon to krypton, the M shell is expanded to contain eighteen electrons while the fourth or N shell is built up to eight electrons. For xenon, the outer shell, the fifth or O shell, contains eight electrons but the penultimate N shell has now eighteen electrons. Radon is formed by changes in three shells: the N shell takes in fourteen more electrons and the O shell, ten electrons, while the sixth or P shell contains eight electrons. Except for helium, each inert gas has an outer shell of eight electrons.

The family is known as the noble, or inert, gases. A group of eight electrons forming a closed shell is called an octet and the electronic configurations shown in Table 3 are considered to be endowed with a special stability. All but the K and L shells may be later expanded to hold 18 (M), 18 or 32 (N) and 18 (O) electrons. The maximum number of electrons in a shell of quantum number n is seen to be $2n^2$ up to and including the fourth shell.

Except for the K shell, each shell is made up of two or more orbits or sub-shells, each energy level being subdivided into a number of closely grouped different levels.

Atomic masses are no longer regarded as fundamental qualities for elements and have been superseded by atomic numbers, which represent not only the order of the element in the periodic table but also the nuclear charge and the number of electrons in the electron shells. Shells are also called energy levels and of course have no material existence themselves.

The Periodic Law

The properties of elements are a periodic function of their atomic numbers. The Periodic Table used today is a long form produced by extending the long periods of the Mendeleeff Table: it is very similar to the Bohr-Thomsen version and inside the back cover. Each horizontal period corresponds to the quantum number assigned to the outer shell which is completed with an octet of electrons (two for the first period) for the inert gas seen at the end of the period. The term, series, as used by Mendeleeff, may now be discarded.

Notations for subgroups In the long form of the Periodic Table, the original designation of A and B subgroups shown in the Mendel6eff Table has been retained. A minority of authors prefer to list transitional elements exclusively as B subgroups and therefore the normal elements as A subgroups. Thus IIIa to VIIa inclusive become IIIb to VIIb respectively and vice versa. However, this is not very important because it is customary to indicate the composition of a family of elements by giving at least the first-named element. Finally, it may be mentioned that some have argued for 18 groups, there being eighteen places (excluding lanthanides and actinides) in the longest periods of the

Periodic Table. Hydrogen has the simplest atom, the most abundant isotope having a nucleus of one proton with one electron outside it. From this simplest of structures, by the addition at each step of one proton to the nucleus and one external electron, with the appropriate number of neutrons to provide its isotopic mass, the atom of any element may be formed in theory.

Chemical character depends on electronic configuration. The K shell may have one or two electrons only. It is expanded during the formation of atoms of the elements hydrogen and helium, the only members of period 1. Hydrogen, which is quite unlike any other element, has sometimes been placed by itself above the table, or even included above fluorine, as well as above lithium, Li.

The lithium atom, which has three electrons, has a completed K shell of two and a new shell, L, starting with one electron. The elements beryllium to fluorine, of atomic numbers 4-9 inclusive, have progressively extra electrons in the L shell until neon is reached. At this element, both K and L shells are full. The M shell commences with sodium, which has eleven electrons, one more than can be accommodated in the K and L shells of its atoms. This shell is built up steadily until argon is reached, all elements being placed in the third period. The electronic configuration of the last member of this period, which will be the inert gas krypton, is 2, 8, 18, 8 and is seen to involve the expansion of the M and N shells.

The energy levels filled in the start of shell N and the further filling of shell M are about the same. The potassium atom contains nineteen electrons, one starting the N shell, the remainder having the argon configuration 2, 8, 8. Two electrons are in the N shell for the calcium atom, otherwise the pattern is that of potassium. Starting with the next element, No. 21, scandium, the penultimate M shell is expanded to contain a further ten electrons while, except for chromium and copper, two electrons remain in the outermost shell. At this stage, it may be assumed that the transference of an electron from the outermost N shell to the penultimate M shell gives a symmetrical distribution, with lower energy, for the five electrons of chromium and ten electrons of copper, which have been added to the penultimate shell. With zinc comes the completion of the M shell and the restoration of two electrons in the N shell. The outermost shell then takes in extra electrons to form the electronic configurations of gallium, germanium, arsenic, selenium, bromine and krypton in succession. Certain metals in Period 4 are called transition metals or transition elements.

Transition elements are characterized by atoms in which an inner electron shell (or level) is in process of expansion but is not yet complete. In this period, by definition, the transition series comprises scandium to nickel inclusive. However, because of some similarities, copper and zinc are sometimes classed with the transition elements in this book they will be referred to as 'associated with transition elements'. Elements other than transition elements and inert gases are called normal elements or representative elements.

In the transition series the absorption of an outermost electron by the penultimate shell occurs more often. By strict definition, the transition series starts with yttrium (Z = 39) and ends with rhodium (Z = 45), although palladium, silver and cadmium may be included.

The first element, caesium (Z = 55) has the electronic configuration of xenon for fifty-four of its electrons and the last electron is placed in the P shell. Similarly, the next element, barium (Z = 56) has two electrons in the P shell. The first member of the transition series, lanthanum, now appears with the expansion of the penultimate shell.

However, a new phenomenon appears with the expansion of the O shell from eighteen electrons in successive stages to thirty-two electrons. The fourteen new elements are called *lanthanons, lanthanides* or *rare earths*. They are very similar to one another. This is a transition series within another transition series, or an inner transition series. Hafnium (Z = 72) is the next element of the transition series which by rigid definition finishes with platinum (Z = 78) followed by gold (Z = 79) and mercury (Z = 80).

The outermost shell is then expanded with the appearance of the sequence, thallium (Z = 81) to radon (Z = 86). As the result of the work of Seaborg and his associates at the University of California, a whole series from Z = 93 onwards has been produced by transmutation starting from uranium. In the best Mendel6eff tradition of prediction, it has been established by these workers that an inner transition series, the actinide series, is formed. This is expected to contain fourteen elements like the lanthanides and so the second member of the outer transition series should be element Z = 104.

Atomic orbitals and quantum numbers

The concept of quantum shells (electron shells) associated with energy levels is based on the interpretation of atomic spectra. However, single shells present too simple a picture to account for the complexities seen. It has already been noted that the simple mechanical model of the atom must be abandoned, the result of Wave (Quantum) Mechanical treatment and application of the Heisenberg Uncertainty (Indeterminacy) Principle. Each quantum shell is replaced by a group of atomic orbitals, except for the first shell which is replaced by one type of orbital.

The atomic orbital represents the boundaries of the space outside the nucleus in which there is the greater probability of finding a particular electron. In the mathematical terms of wave mechanics, the wave functions describing single electrons are called orbitals. Each orbital may hold only two electrons, each spinning in the opposite sense to the other. Electrons are assigned to orbitals so that those associated with the lowest energy levels are filled first. For orbitals of a given type as many single electrons as possible are accommodated before pairing starts. The number of unpaired electrons determines the valency.

Orbitals of a given type occur in sets in successive energy levels:

1*s* 2*s* 3*s* 4*s* ...

2*p* 3*p* 4*p* ...

3*d* 4*d* ...

4*f* ...

The number refers to the Principal Quantum Number (n) already mentioned, the energy level nearest to the nucleus (K shell) having $n = 1$. The letters refer to the Subsidiary Quantum Number (l) which serves to distinguish the types of orbital in each energy level. The letters originate in the spectroscopic terms for the series: sharp, principal, diffuse, fundamental. For a given value of n, the subsidiary quantum number may assume values up to $(n - 1)$:

1	0	1	2	3 ... (n - 1)
	or	or	or	or
	s	*p*	*d*	*f*

The subsidiary quantum number is also called the Angular Momentum Quantum Number. The first energy level (n = 1) has only one type of orbital (*s*), the second, two types (*s*, *p*), the third, three (*s*, *p*, *d*), and so on.

The p, orbital differs from the s orbital in the same energy level by having an energy distribution which depends on the angle and radial distance from the nucleus: it has a directional character. By symmetry, the three *p* orbitals in a given level have the same energy, unless an electric or magnetic field is applied to the atom. In a powerful magnetic field the lines of an atomic spectrum are split up, the new lines being associated with only slightly different energies from those seen in the absence of the field. This is the Zeeman Effect.

A similar happening in an electric field is called the Stark Effect. The application of a strong magnetic field leads to different energies being associated with orbitals which were previously identical. The magnetic quantum number (m,) is assigned to distinguish between orbitals of the same type in a given level. It may assume any integral value between +1 and -1 including zero:

l	*m*
0(≡*s*)	0
1 (≡*p*)	+1, 0, -1
2(≡*d*)	+2, +1, 0, -1, -2
3(≡*f*)	+3, +2, +1, 0, -1, -2, -3

The fourth energy level (n = 4) comprises one s orbital, three p orbitals, five d orbitals and seven f orbitals. The three quantum numbers given are found in the Schrodinger wave equation, which has only been precisely solved for hydrogen. A pictorial representation of orbitals, or rather, the electron-cloud probability pattern they represent, may be attempted.

The Is orbital is spherically symmetrical about the nucleus, rather like a fluffy ball of wool, though with no material existence, thinning out at the periphery as the probability of finding the electron rapidly diminishes. The 2*s* orbital is similarly spherically symmetrical but less dense and at a greater distance and separated from the Is orbital by a region (spherical nodal surface) where the probability of finding the electron is low.

In the higher quantum states, the electron-cloud spreads over a much larger region of space, and since the probability refers to finding one electron only, the density of the cloud is much reduced. There are three 2*p* orbitals, 'dumb bell' in shape placed along axes at right angles to each other. The z axis is defined by reference to an external magnetic field and the x and y axes by reference to additional external fields, as created (for example) by adjacent atoms in a crystal. Notice the nodal plane (virtual zero probability of locating an electron) at the origin, passing through the atomic nucleus.

Each orbital may contain two electrons. A fourth quantum number, the Spin Quantum Number (m_s) is assigned to distinguish between these electrons. From studies of the fine structure of atomic structure, it is considered that electrons behave as if spinning about their centres. An electron is given a value of m_s equal to +½ or -½. The difference between the energies of the electrons which differ only in ms is very small, but increases with atomic number. The state of an electron is defined by its four quantum numbers, no two electrons in the same atom having exactly the same four quantum numbers. This is the Pauli Exclusion Principle and was put forward empirically. When the electrons in an atom are in the lowest possible energy levels consistent with this principle, the atom is said to be in its ground state.

The Rule of Maximum Multiplicity states that in

filling a *p*, *d* or *f* orbital, as many unpaired electrons as possible are placed before coupling of electrons with opposed spins is allowed. Electrons are placed in orbitals of the lowest energy level and orbitals filled in order of increasing energy, the inert gas at the end of the period resulting from completion of the orbitals being shown in parentheses.

However, energy levels change relative to each other with change in atomic number, while above atomic number 60, overlap of principal shells no longer occurs. The completion of successive energy levels may be considered further. Numbers below the coupled electrons refer to the magnetic quantum numbers.

The other numbers are principal quantum numbers and the letters, subsidiary numbers, (*s*≡O, *p*≡1, *d*≡2, *f*≡3). Spin quantum numbers (±½) have not been shown. The electrons in the 3p orbitals have (for example) the following quantum numbers:

3, 1, +1, +½	3, 1, 0, +½	3, 1, -1, +½
3, 1, +1, -½	3, 1, 0, -½	3, 1, -1, -½

The numbers 2, 8, 18 are the numbers of elements in successive periods of the periodic table. By use of a small subscript number to represent the number of electrons in the orbital, or group of similar orbitals, indicated, electronic configurations may be listed:

H	$(1s)^1$
He	$(1s)^2$
Li	$(1s)^2(2s)^1$
Be	$(1s)^2(2s)^2$
B	$(1s)^2(2s)^2(2p)^1$
C	$(1s)^2(2s)^2(2p)^2$

N	$(1s)^2(2s)^2(2p)^3$
O	$(1s)^2(2s)^2(2p)^4$
F	$(1s)^2(2s)^2(2p)^5$
Ne	$(1s)^2(2s)^2(2p)^6$

This building-up is sometimes called the *aufhau* process. Argon, the next inert gas, has the electronic configuration:

$$\text{Ar} \quad (1s)^2(2s)^2\,(2p)^6(3s)^2(3p)^6$$

In the next period a transition series starts with scandium and ends with nickel. This corresponds to the appearance of 3d orbitals:

	K	$(1s)^2(2s)^2(2p)^6(3s)^2(3p)^6$	$(4s)^1$
	Ca		$(4s)^2$
Transition	Sc	Argon core	$(4s)^2(3d)^1$
series	Ti		$(4s)^2(3d)^2$
	V		$(4s)^2(3d)^3$
	Cr		$(4s)^1(3d)^5$
	Mn		$(4s)^2(3d)^5$
	Fe		$(4s)^2(3d)^6$
	Co		$(4s)^2(3d)^7$
	Ni		$(4s)^2(3d)^8$
	Cu		$(4s)^1(3d)^{10}$
	Zn		$(4s)^2(3d)^{10}$

Krypton, the next inert gas, has the electronic configuration:

$$\text{Kr} \quad (1s)^2(2s)^2(2p)^6(3s)^2(3p)^6(4s)^2(3\text{d})^{10}(4\text{p})^6$$

with 36 electrons in all. Returning to the second period, the $2p$ electrons are shared between $2p_x$, $2p_y$, $2p_z$ orbitals, singly at first, and then in pairs:

C $(1s)^2(2s)^2(2p_x)^1(2p_y)^1$

N $(1s)^2(2s)^2(2p_x)^1(2p_y)^1(2p_z)^1$

O $(1s)^2(2s)^2(2p_x)^2(2p_y)^1(2p_z)^1$

F $(1s)^2(2s)^2(2p_x)^2(2p_y)^2(2p_z)^1$

Ne $(1s)^2(2s)^2(2p_x)^2(2p_y)^2(2p_z)^2$

A mutually paired set of electrons in an outer shell is called a *lone pair*, e.g. (2*s*)2 and (2p,)2 for oxygen, (2s)2 for nitrogen, etc. The number of singly occupied orbitals determines the valency of an element. Nitrogen is tervalent:

	$1s$	$2s$	$2p_x$	$2p_y$	$2p_z$
N	↑↓	↑↓	↑	↑	↑

When electrons are transferred to form ions or shared to form covalent bonds, electrons are coupled in pairs with opposing spin.

The 's, p, d, f blocks' of elements

The modern Periodic Table reflects the order of energy levels in the atoms of elements. Sometimes elements are

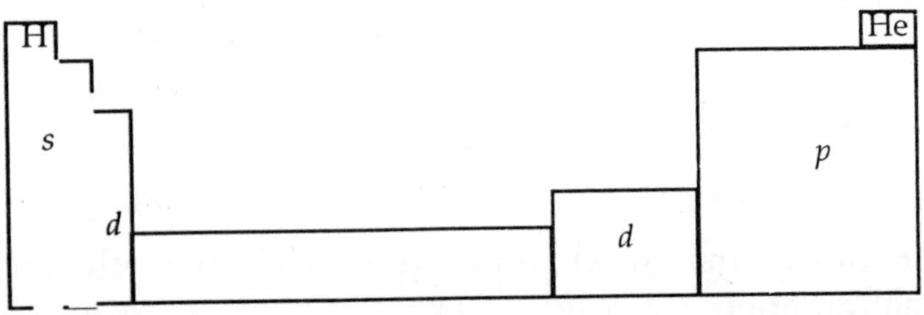

Fig. 3. Orbitals and the periodic Table: s, p, d, f blocks

described as belonging to a particular 'block' according to whether *s, p, d,* or *f* levels are being filled, and the number of elements in a period will depend on the number of 'blocks' traversed, each holding sets of 2, 6, 10 and 14 corresponding to *s*, *p*, *d* and *f* orbitals. The d block is split by the *f* block, and helium, $(1s)^2$, is isolated on the right.

Quantitative evidence

The amount of energy required to eject one electron from an isolated atom of an element, leaving both the resulting ion and electron without surplus kinetic energy, is called the first ionization energy (or ionization potential) of the element. Successive removal of electrons requires input of the 1st, 2nd, 3rd, ... ionization energies. Energy is always required to draw the electron away from the nuclear charge of the atom or ion and is a measure of the bonding energy of the electron removed.

The value is always positive. There is always a surge in value after removal of the valency electrons, as shown by the 3rd ionization energy in the table. The first ionization energy may be measured by the potential difference required to give electrons, which are being used to bombard gaseous atoms, enough energy to eject the first electron from these atoms. Prior to the ejection of the electrons, new spectral lines appear as electrons return to lower energy levels (shells) within the atoms after promotion to higher levels (outer shells). As the potential is raised, ionization is reached.

The first ionization energy is therefore a measure of the energy required to remove the first electron from an isolated atom (in the gaseous state). It is a periodic property of atoms for elements 1-56 (hydrogen to barium) inclusive. Ionization energy increases along a period with increasing proton number because the effective nuclear charge increases while electrons are being accommodated in the same quantum shell i.e. at the same energy level; it decreases down the groups because electrons in outer shells tend to be shielded more from the nucleus.

Mutual repulsion between electrons of opposite spin in the same quantum level accounts for minor dips in the curve. Values for ionization energies are given with individual group studies later. Close examination of the curve provides evidence for the arrangements of electrons known as sub-shells or orbitals. In building up atoms from lithium to neon, the first peak corresponds to completion of the 2*s* orbital, the next illustrates that the incoming 2*p* electrons remain unpaired until four electrons have to be accommodated. Finally, the stability of the noble gas electronic configuration should be noted. The ns^2np^6 ($1s^2$ for He) configurations are the most stable in the Periodic Classification, all requiring a high input of energy for removal of an electron.

	1s	*2s*	*2p*			*Configuration*
Li	↑↓	↑				$(1s)^2(2s)^1$
Be	↑↓	↑↓				$(1s)^2(2s)^2$
B	↑↓	↑↓	↑			$(1s)^2(2s)^2(2p_x)^1$
C	↑↓	↑↓	↑	↑		$(1s)^2(2s)^2(2p_x)^1(2p_y)^1$
N	↑↓	↑↓	↑	↑	↑	$(1s)^2(2s)^2(2p_x)^1(2p_y)^1(2p_z)^1$
O	↑↓	↑↓	↑↓	↑	↑	$(1s)^2(2s)^2(2p_x)^2(2p_y)^1(2p_z)^1$
F	↑↓	↑↓	↑↓	↑↓	↑	$(1s)^2(2s)^2(2p_x)^2(2p_y)^2(2p_z)^1$
Ne	↑↓	↑↓	↑↓	↑↓	↑↓	$(1s)^2(2s)^2(2p_x)^2(2p_y)^2(2p_z)^2$

In the later periods it will be seen that the d^5 and f^7 half-filled shell configurations have enhanced stabilities similar to that for p^3 above, an illustration of Hund's Rule of maximum multiplicity and the idea of exchange energy.

Successive ionization energies

Further evidence for the orderly build up of the electronic configuration of atoms comes from a study of the energy required to remove electrons one by one.

When the measured values are plotted against the order of ionization, or ionization number which equals the number of electrons removed, this is apparent. However, because of the wide range of values of ionization energies it is usual to plot the logarithm of successive ionization energies against ionization numbers, i.e. expressing ionization energies as powers of 10 and plotting the index e.g. instead of (say) plotting 1000, this is expressed as 10^3 and plotted as 3; 100 000 would be plotted as 5, rather like pH plots except that here minus the logarithm is used to convert negative indices into positive numbers.

Each sequence of points lies on a rising curve showing that successively larger ionization energies are required to eject an electron from the atom. As each electron goes, so some screening is lost and the positive nuclear charge exerts a more powerful attraction for the electrons left. Steep rises between certain successive (logarithms of) ionization, energies are seen, showing that proportionately more energy is suddenly required to shift an electron. This electron was located nearer to the nucleus than the previous one because the effect is more pronounced than what might have been expected from a progressive rise in effective proton number.

The broken vertical lines therefore denote a change in quantum shell. In the light of what has been discussed before, for potassium the first electron ejected has come from the 4s orbital and not from the 3d orbital. Note the disposition of the curves: because of its higher effective nuclear charge for a given number of residual electrons, the curve for potassium lies above that for sulphur, and so on.

Exchange energy

This concept comes from quantum mechanics and arises from the impossibility of distinguishing between electrons during the interaction of atoms. It is a function of the number of pairs of electrons with parallel spins (when clockwise and anticlockwise spins of electron pairs, ↑↓ are treated independently) and different from the energy associated with the electrostatic interaction between nucleus and electrons, and between electrons. In the mathematical study of the hydrogen molecule, H_2, the total energy for the molecule appears to be lowered by allowing either electron to be near to either nucleus, the electrons being identical and subject to simultaneous attraction to the two positive nuclei (instead of one, as in the atom). Each electron is assumed to have a separate 'identity' so that

↑↑ = 2 electrons with parallel spins: 1 pair

↑↑↑ = 3 electrons with parallel spins: 3 pairs

↑↑↑ = 4 electrons with parallel spins: 6 pairs

↑↑↑↑ = 5 electrons with parallel spins: 10 pairs

Thus ↑↑↑↓↓ depicts 5 electrons with 3 + 1 = 4 pairs with parallel spins. This is the basis of Hund's Rule that in orbitals of equal energy electrons are assigned with parallel spins until coupling is unavoidable. As an illustration the number of pairs of electrons with parallel spins may be compared for maximum pairing and minimum pairing. The exchange energy (directly dependent on the number of parallel spins) is greatest for p^2, p^3, p^4 configurations when Hund's Rule is obeyed. Without attempting to explore the mathematical treatment, we may accept that an extra stability is conferred on an atom where the exchange energy factor appears.

The gain in exchange energy more than compensates for the loss due to transferring an electron between energy levels, from the 4*s* orbital to the 3*d* orbital, i.e. $(4s)^1(3d)^{10}$ is preferred to $(4s)^2(3d)^9$. The term exchange energy must not be associated with an oscillation of electrons or any other picture. It arises from the mathematical equations of quantum mechanics and represents an extra stabilization factor in studying the energy of atoms, with no classical analogy to be drawn upon. Indeed, if we do not know which electron is which, to say that they change places is meaningless. By suitable compression this form of the Periodic Classification may be converted into the Mendeleeff Table.

However, the creation of the subgroups leads to paradoxes such as the inclusion of Cu (copper), Ag (silver) and Au (gold) in juxtaposition to K (potassium), Rb (rubidium) and Cs (caesium), although these elements are quite dissimilar: further, Mn (manganese) and Br (bromine) provide another example of unlike elements placed together. The very real success of the Mendeléeff Table in other ways leads the unwary to seek resemblances where none exist.

The long form avoids this. It shows the grouping of similar elements more clearly, both in the same vertical families and in the long period transition series. The gradation in properties and trends across the table is also more obvious. Groups Ia and Ib are widely separated in the table which is now in current use. The elements of Group la are a well-defined family named the alkali metals: lithium, sodium, potassium, rubidium and caesium.

Melting-points range from 186°C for lithium down to 30°C for caesium which has the highest atomic number. No data are available for francium but there is no doubt about its belonging to the alkali metals family. Of low density and soft enough to be easily cut with a knife, these metals are good conductors of electricity and fairly good conductors of heat, are soon corroded in air and react vigorously with water forming hydrogen and the metal hydroxide. With a dilute acid, the liberation of hydrogen proceeds with violence and the resulting salts are very similar to each other. With no physical or chemical resemblance to Group la are the so-called coinage metals of Group Ib: copper, silver and gold. Excellent conductors of electricity and heat, they are hard and show a high lustre with surface tarnishing for copper and silver. They have high melting-points: copper (1083°C), silver (961°C) and gold (1063°C), and have high densities.

These metals do not react with water or non-oxidizing acids. The valency shown by Group la is I while one valency state of Group Ib is also 1. This is the only similarity. Although by strict definition the elements of Group Ib are not transition elements it is often instructive to compare their characteristics with those of the true transition series. This is in accordance with Mendeleeff's observations.

Elements of the transition series are all metals. In varying degrees they show the ability to combine in various valency (and oxidation) states, form coloured ions and complex compounds readily and frequently, as elements or compounds and behave as catalysts. Corresponding salts sometimes form isomorphous crystals. Quite different from the alkali metals except in the degree to which they hold their properties are the elements of Group VIIb: fluorine, chlorine, bromine, iodine and astatine (the halogens).

These elements are all strongly reactive non-metals although there are little data available about the short-lived radioactive astatine. Non-conductors of electricity, they have melting-points ranging from -223°C for fluorine to +114°C for iodine, and boiling-points from -118°C for fluorine to +184°C for iodine. They form very similar compounds, with a well-marked gradation in reactivity, fluorine being the most reactive non-metal known. Each member of the halogens is separated from the nearest alkali metal by an inert gas. The great value of the Periodic Table results from the way in which the detailed memorization of facts may be reduced. By the grouping of elements to show similarities, contrasts and trends in characteristics, a wider grasp and deep understanding of the subject may be attained.

Although the properties of an element and its compounds may be generally predicted from a knowledge of the behaviour of nearby elements in the table there are pitfalls: some elements do show anomalous properties, chemical and physical, and predictions should be checked by use of suitable reference books or, if feasible, by direct experiment. The most useful form of the Periodic Table, designed for ease of reference, will be two-dimensional and the long form is the simplest: electrons are systematically added in successive steps along any period working from left to right and elements fall into vertical family groups.

Admittedly, hydrogen is by itself and for this reason is sometimes included twice, above the alkali metals (as here) and above the halogens, but no geometrical solution will show the unique nature of hydrogen unless it is set aside by itself! There seems to be no simple way in which the lanthanides and actinides can be directly included in the table. Splitting the (outer) transition series so as to include these elements would reduce the value of the rest of the table because it would become too wide and cumbersome.

The gradation of properties shown in using the Periodic Classification will be considered further in the next chapter, which deals with the subject of valency, with which chemical nature is firmly linked.

Transuranium elements

In early days, the elements thorium, protactinium and uranium were usually positioned below hafnium, tantalum and tungsten respectively in the Periodic Table. The synthesis of neptunium, a new element of atomic number 93, by McMillan and Abelson in 1940 altered the position radically. Neptunium was discovered during the investigation of the fission products (i.e. breakdown products) of uranium induced by neutrons. $^{239}_{93}Np$ results from the β decay of $^{239}_{92}U$ formed by neutron capture in $^{238}_{92}U$. Now the position immediately to the right of tungsten is occupied by rhenium. The properties of neptunium resembled those of uranium rather than rhenium, the implication being that a new transition series was being built up. This is called the actinide series and is shown at the bottom of the Periodic Table.

The preliminary studies of the new element were done by radiotrace techniques on quantities of the order of 10^{-10} g, the radioactive character of the atoms being studied. Thorium, protactinium and uranium are no longer placed below hafnium, tantalum and tungsten. Later in 1940, McMillan, Kennedy, Wahl and Seaborg bombarded uranium with deuterons and produced $^{238}_{93}NP$, another isotope of neptunium, which decayed to a new element, plutonium $^{238}_{94}Pu$. Its military potential led to its production in 'visible quantity'.

Cunningham and Werner prepared about a microgram of 94 in 1943. Ultra-micro chemical techniques in which volumes of liquids in the range of 10^{-1} to 10^{-5} cm^3 were measured to within 1076 were developed and chemical work was performed under the microscope. Using atomic (or nuclear) reactors, plutonium has been produced in kilogram quantities.

The synthesis of other M (24 elements followed rapidly. Americium ($^{241}_{95}Am$) and curium ($^{242}_{96}Cm$) were identified largely on the basis of prediction of properties from scrutiny of the Periodic Table. The synthesis of elements beyond lawrencium should see the start of an outer transition series: rutherfordium (104) (under hafnium), hahnium (105) (under tantalum), 106 (under tungsten) and the end of the period with an element of inert gas configuration (118). In writing about his work in this field Seaborg predicted that element 102 would have a stable tervalent state and somewhat unstable bivalent state, with an unusually low density and relatively high volatility.

Element 103 was expected to be only tervalent. Element 104 would probably be quadrivalent in solution only and resemble hafnium of the previous period. Element 105 should resemble niobium and tantalum. Element 106 should fit into the Cr, Mo and W sequence with valencies 3, 4, 5 and 6. The synthesis of further elements is expected to become increasingly difficult because the predicted half-lives of the new isotopes are expected to be very short, of the order of seconds or minutes although there could be a sequence around elements 114 and 126 with somewhat higher stability.

Electronic and Magnetic Compounds

The study of the electronic and magnetic properties of inorganic compound has yielded enormous amount of information, and for the transition metals has dominated the development of bonding theories; it is useful to present a chapter to relating these observations to the theories proposed.

Electronic spectra

It is usual to study spectra in the range 200 nm to 1000 nm, the near ultra-violet to the near infra-red. These energies correspond to electronic transitions between weakly bonding and weakly antibonding, bonding and non-bonding, and nonbonding and antibonding levels. At high energies almost all compounds are opaque. At lower energies, molecular vibrations complicate the spectra. Many transition metal complexes show absorptions at relatively low energies for electronic transitions. These are associated with transitions between d orbitals whose degeneracy has been removed by air environment of lower than spherical symmetry. For example, in an octahedral complex, transitions involving electron, jumps from the t_{2g} orbitals to the e_g orbitals.

These orbitals are often only weakly bonding or antibonding—their energy separation is small. We may use the absorption energies to measure the ligand field splitting energies Δ, but it should be noted that only in a few cases does the absorption energy correspond exactly to Δ. The intensity of a transition is proportional to tire transition probability $|<\psi_{excited}|\mu|\psi$ ground $>|2$, and the symmetries of ψ_{ex}, ψ_{gd} and μ may therefore indicate if the transition will have zero intensity. Electronic transitions are almost invariably caused by interaction with the oscillating electric dipole moment of electromagnetic radiation; there are perhaps half a dozen exceptions where electric dipole forbidden transitions acquire visible intensity by other mechanisms.

The electric dipole moment operator μ transforms as the quantities x, y, and z, corresponding to the t1U I.R. in octahedral symmetry, and the p I.R. in spherical symmetry. If we consider a transition of the kind $t_{2g}{}^1 \rightarrow e_g{}^1$ then ψ_{ex} and ψ_{gd} are both g type $< \psi_{ex}|\mu|\ \psi_{gd} >$ has g x u x g = u type symmetry and consequently is zero. The transition is forbidden, yet $3d^1$ complexes such as $[Ti(OH_2)6]^{3+}$ are quite strongly coloured.

This is caused by the excitation of vibrations in the complex by the radiation at the same time as the electronic excitation: if we simultaneously excite a vibration from its g type groundstate to a u-type excited state, the total wave function ψ_{ex} will now be u-type and the product will have u X 11 x 9 = 9 type symmetry, and the transition moment will not vanish. This co-excitation of electronic and vibrational levels leading to an apparent breakdown of a selection rule is known as a *vibronic* transition. It re quires that the orbitals involved in the transition should be 'coupled' with, or affected by the molecular vibrations, i.e. that they should not be completely non-bonding, and results in the absorption spectra being broadened by the vibrational energy taken up or released.

These spectra are consequently much broader than the sharp single lines of atomic spectra where the transitions are pure electronic and rigorously obey the selection rule $\Delta l = \pm 1$. In compounds of the lanthanide elements, the 4f orbitals are almost completely non-bonding, and spectra resulting from f-f transitions are much weaker and narrower than their d-d counterparts. The selection rule $\Delta S = 0$ is much more rigorously obeyed, as it is not affected by any vibronic coupling. Thus complexes with a P high spin configuration ($t_{2g}3\ e_g{}^2$), which cannot undergo a $t_{2g} \rightarrow e_g$ transition without pairing an electron spin in the e_g levels, and consequently a change in *S*, have very weak d-d absorption spectra—for example the very pale pink manganous Mn^{2+} salts.

The reason that they are seen at all is the result of a spin-orbit coupling linking the pure atomic quantum numbers L and *S*—if *S* is no longer a pure quantum number (i.e. the Russell-Saunders coupling scheme is not 100 percent accurate) then the $\Delta S = 0$ rule may be broken. The secondary effects which allow transitions to be seen, are not however powerful enough to give the absorption bands great intensity. Thus if we measure the strength of the transition by the molar extinction coefficient, typical values for the various transitions mechanisms are:

Fully allowed electric dipole 10^5
Vibronically allowed 10^2
$\Delta S \neq 0$, but spin-orbit allowed < 1

In tetrahedral symmetry, three of the d orbitals belong to the same I.R. as the p orbitals (t2), and some d-p mixing can occur. This results in a considerable increase in intensity compared with the pure d-d transitions of octahedral complexes: the d-d transitions of tetrahedral $[COCl_4]^{2-}$ have extinction coefficients nearly a hundred times higher than octahedral $[Co(OH_2)_6]^{2+}$. In the discussion of atomic spectra it was emphasised that a partially filled shell had different electronic energies depending on the relative arrangements of the electrons, and that these energy states could be described by Russell-Saunders terms if the spin-orbit coupling was not too great.

In the interpretation of the spectra of complexes with partially filled d shells we would expect such electron repulsion effects to be important. We may discuss these effects from one of two extreme positions - 1. the electron repulsion effects much greater than the ligand field effects (the weak field case); 2. the ligand field effects much stronger than the electron repulsion effects (the strong field case). This is treated by taking the Russell-Saunders terms and applying the ligand field splitting as a small perturbation. For the dl case there are no electronic repulsion effects, but for d^2, the Russell-Saunders terms are ^3F, ^3p (spins parallel); ^1S, ^1D, ^1G (spins paired). Hund's rules would lead us to predict that ^3F is the lowest term.

The energy differences between terms of the same multiplicity may be expressed in terms of Racah's parameters B and C. In this case, the difference between the groundstate and the ^3p state is 15B. If we now consider the effects of lowering the symmetry to octahedral, the ^3F

term in spherical symmetry transforms as $^3A^{2g}$ + $^3T_{2g}$ + $^3T_{1g}$, while the ^3p transforms as $^3T_{1g}$. This is very similar to the I.R.s spanned by f orbitals (a_{2u} + t_{1u} + t_{2u}) and p orbitals (t_{1u}) respectively but the terms are g-type since they arise from the g-type d orbitals. As with perturbations in general, the octahedral ligand field causes a splitting of the 21-fold degenerate ^3F term, in this case into three different terms. It can be shown readily for the ^3F terms, that as Δ increases $^3A_{2g}$ rises in energy, $^3T_{1g}$ falls, and $^3T_{2g}$ rises slowly.

The ^3P term, now $^3T_{1g}$ (written $^3T_{2g}$ (P) to distinguish it from the other $^3T_{1g}$ (F)), rises slowly in energy. We may plot the variation in energy of the terms, against increasing Δ as in Fig. 4. We would thus expect three absorptions arising from transitions from he $^3T_{1g}$ (F) ground state to the $^3T_{2g}$, $^3A_{2g}$, and $^3T_{1g}$ (P) levels.

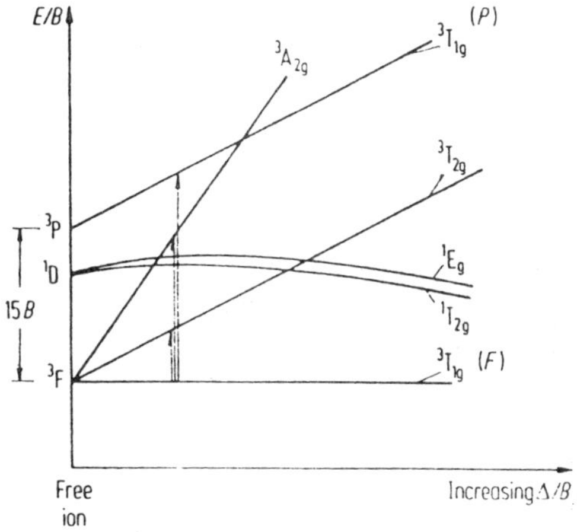

Fig. 4. Tanable Sugano diagram showing the splitting of Russell-Saunders terms relative to the groundstate $^3T_{1g}$ with increasing ligand field.

If we consider an example $[V(H_2O)_6]^{3+}$, only the first two transitions can be observed, the third high energy band being masked by a more intense charge transfer band (q.v.). The problem of masking by other bands is quite common. There is a ^1D level lying between the ^3F and ^3P levels, but any transition to this level (^1E_g and $^1T_{2g}$ in the ligand field) is spin forbidden, and the transitions are masked by the spin allowed transitions, some of which are, in turn, masked by the dipole moment allowed charge transfer bands. It should be noted that the strength of the interelectronic repulsion is such that one cannot generally assign a precise number of electrons to the e_g or t_{2g}, (i.e. an $e_g^a t_{2g}^b$ configuration) but must consider the ligand field as changing the energy and lifting the degeneracy of well-defined Russell-Saunders terms.

Derivation of parameters from the spectra

In the weak field case, the ligand field is applied as a perturbation. The calculations are relatively straightforward. The effects are i) to lift the degeneracy of the Russell-Saunders terms, e.g. ^3F becomes ^3T1g + $^3T_{2g}$ + $^3A_{2g}$ —this splitting is dependent only on the magnitude of Δ, and ii) as Δ increases, to cause mixing of levels of the same symmetry but originating in different Russell-Saunders terms.

Thus in Fig. 1, as Δ increases, a mixing of the $^3T_{1g}$ (F) and $^3T_{1g}$ (P) terms is seen. As one would expect from perturbation theory the mixing is dependent on the value of Δ, *and* the initial separation of ^3F and ^3p. It is possible from analysis of the spectra to obtain values of Δ, and the Racah parameter B. It is found that the value of B is always less than that known for the equivalent free metal ion (i.e. B for $[Ni(NH_3)_6]^{2+}$ is less than B for gaseous Ni^{2+}). There is, of course, a logical flaw in the use of the Racah parameter to discuss electron repulsion in complexes, since it was introduced to describe spherical symmetry, while the symmetry of the complex is lower.

The continued use of B is thus an artificial averaging of the electron repulsion effects in the eg and t_{2g} orbitals which we have previously shown to be differently affected by chemical bonding. The strong field treatment avoids this problem by introducing a larger number of electron repulsion parameters: nine are needed for a full description, but a reasonable description is given with three. Such experimental results that are available show that there is an appreciable difference in the electron repulsion between a pair of electrons in the e_g subshell, and between a pair in the t_{2g} subshell. (i.e. the nephelauxetic effect varies).

However, there are not always enough absorption bands in the spectrum of the complex to obtain all the new electron repulsion parameters, especially for strong field complexes, where the large value of Δ results in many d-d transitions having similar energies to charge transfer bands, and being obscured. In view of the information on chemical bonding that d-d spectra may yield, a considerable amount of effort has been spent in calculation of energy levels.

The first approach assumed that the splitting of the d levels arose solely from the electrostatic potential of negative ions surrounding the metal ion. Although this purely electrostatic "crystal field" theory worked quite well for the highly ionic species to which it was initially applied, it became clear as more experimental results appeared that it had several shortcomings. The qualitatively more accurate molecular orbital theory was much harder to use quantitatively: quite acceptable atomic wave-functions for the d-orbitals are available for crystal field theory, but molecular orbitals are much harder to calculate. At the moment the most acceptable approach would seem to be the *angular overlap* model of Jørgensen and Schäffer which enables calculations to be made on a M.O. basis. For further details the reader should consult Gerloch and Slade.

The problem of calculating the pattern of energy levels is simplified by the fact that the splitting depends mainly on the symmetry, and thus the pattern is not too much affected by the physical cause we choose to give the splitting. Diagrams similar to Fig. 1 are given in many introductions to transition metal chemistry, for all numbers of d electrons d_1-d^{10}. These diagrams either assume fixed values of the Racah parameters B and C or assume a fixed ratio B/C and plot E/B against Δ/B. Most of the information gained from spectra comes from the variation of the ligand field splitting Δ, and the Racah parameter B, and we now discuss their variation. For a given metal, the order of ligands producing increasing A is almost invariably the same; for a given ligand the order of metal ions giving increasing A is also pretty well constant.

These two series are referred to as the *spectro-chemical series* for ligands and central ions, Jorgensen has pointed out that the ligand field splitting may be factorised into a ligand and a metal component: Δ = f.g where *f* is a ligand and *g* a metal parameter; *f* is arbitrarily put to 1.00 for water. Those species such as CN^- and NO_2^- which have low-lying *π* antibonding orbitals which may accept electrons from the t_{2g} subshell, and which will thus increase Δ, have high *f* values.

Those ligands which may act as p donors (halogens) have rather low *f* values. It was initially thought that the value of *f* was determined only by the atom of the ligand actually bound to the metal; however sulphur can vary from quite low (SCN^-, $(C_2H_5O)_2PS_2^-$), to quite high (CH_3-S-CH_2-CH_2-S-CH_3), so next-nearest atom effects can be important. The interpretation of *f* values is complicated by the fact that a low value may be due to strong *π* donation by the ligand, or simply weak overlap, while high values may indicate strong *σ* donation (probably the case for NH_3) or a mixture of *σ* and *π* effects (for example NO_2^-).

The metal series is rather more interesting. As the oxidation state (and formal charge) of the metal atom increases, g rises considerably: Co(II) 9, Co(III) 18.2; Ni(II) 8,7, Ni(IV) 22. This is not easy to explain on a pure ionic model as increasing charge would be expected to shrink the d orbitals and thus decrease their interaction with the ligands. A M.O. approach would predict a greater donation of charge to the metal as the oxidation state increases as well as an increased overlap, leading to stronger bonding effects, and consequently greater Δ. It is found experimentally that pressure increases A, and that metal-ligand bond lengths decrease with increasing oxidation state of the metal. It should be noted also that *g* for 2nd and 3rd row transition elements is much greater than for their 1st row analogues:

Co(III) 18.2, Rh(III) 27.0, Ir(III) 32; Cr(III) 17,4, Mo(III) 24.6.

Δ values are always much greater for 2nd and 3rd row complexes which are almost exclusively low spin. This can again be explained by the greater overlap of the 4d and 5 d orbitals with ligands, as a result of the more diffuse character of orbitals with one and two nodes.

The *nephelauxetic effect* is generally characterised by a number $\beta = B/B_0$ where B is the Racah parameter obtained from the spectrum of the complex and B_0 is the Racah parameter of the free gaseous ion, Jorgensen has shown that this also may be factorised in the form $(1-\beta)$ = h (ligand) - k (metal).

For ligands the series is very different from the spectrochemical series:

$F^- < H_2O < NH_3 <$ en $< ox^{2-} < NCS^- < Cl^- < CN^- < Br^- < I^- < (C_2H_5O)_2PS_2^-$

This series corresponds to decreasing electronegativity

and increasing reducing power (the capacity to lose electrons). It also corresponds to an increasing overlap with the metal, implying that at least part of the effect arises from the mixing of ligand and metal orbitals. For metals, the series is less well defined, and is quite closely related to the spectrochemical series; however, there is no sharp increase in *k* values on moving from 3d to 4d and 5d transition metals. There is in fact a slight decrease from Co(III) to Ir(III).

The nephelauxetic effect increases with the oxidation state of the central ion. There are two effects commonly used to explain the nephelauxetic effect, central field effects, and symmetry restricted covalency. Central field effects are based on the fact that the transfer of negative charge from the ligands to the metal, and the negative potential of the ligands acts to reduce the effective charge on the metal atom.

The lower the effective charge, the less strongly the d electrons are drawn together, and thus the weaker the repulsion between them. Symmetry restricted covalency reflects the formation of M.O.s by the e_g and t_{2g} d-orbitals—this involves a spreading out of the one electron orbital over the ligands with a consequent decrease in the inter-electron repulsion.

Since the eg and t_{2g} will form different M.O.s, one would expect them to show different nephelauxetic effects as a result of symmetry restricted covalency—this is shown by the three different electron repulsion parameters discussed before.

Jahn- Teller effect

Jahn and Teller proved a theorem which requires that "any non-linear system where there is orbital degeneracy has a vibrational mode which will lower the symmetry of the system and also lower the energy of the system". Thus any octahedral complex which has orbital degeneracy—that is, one in which the ground state is not of A or B type symmetry—will be expected to distort to lower its symmetry and give a more stable system. As an example consider a complex with a full t_{2g} subshell and one electron in the e_g shell. This has the groundstate ^2E, which is degenerate, and may be regarded as involving either occupation of the d_z2 or the $d_{x^2-y^2}$ orbital.

There exists a vibration which involves the simultaneous closer approach of the two ligands along the z axis, and the moving outwards of the 4 ligands along the x + y axes. This will increase the overlap of the z axis ligands, and decrease that of the x and y axis ligands. The d_z2 orbitals thus becomes more antibonding and the $d_{x^2-y^2}$ less so. This corresponds to a lifting of the degeneracy of the ^2E groundstate. If the single electron of the species is in the $d_{x^2-y^2}$ orbital, the energy of the complex will be lowered and the distortion favoured in accordance with the Jahn-Teller theorem.

The theorem is general, and also qualitative, not predicting the magnitude of the distortion. For a large distortion the orbital energy should be sensitive to changes in the metal ligand bond length—i.e. it should be strongly bonding or antibonding. Distortions due to the Jahn-Teller effect are normally greater when the degeneracy is in the occupation of the e_g levels than the t_{2g} levels of an octahedral complex. If the t_{2g} levels are involved in bonding, for example in $V(CO)_6(3d^5)$, groundstate $^2T_{2g}$, the structure is distorted, the t_{2g} levels being bonding as a result of p donation to the carbonyl groups. In complexes of the lanthanides, the 4f orbitals are almost completely non-bonding, and show orbital degeneracy, but no Jahn-Teller effect.

If the Jahn-Teller distortion energy is sufficiently great, the distortion shows up in the metal-ligand bond lengths. This is particularly common in the case of Cr^{2+} (high spin $3d^4$) and Cu^{2+} ($3d^9$) compounds where there is a degeneracy arising from one 'hole' in the eg levels. CrF_2 and CuF_2 both show approximate octahedral symmetry but the bond lengths are not equal:

CuF_2	4F at 1.93 Å	2 at 2.27 Å
CrF_2	4F at 2.00 Å	2 at 2.43 Å

The 4 short and 2 long bonds distortion is the most common, but there are other modes (K_2CuF_4 shows 2 short and 4 long bonds). The effect is also seen in d-d spectra. $[Ti(H_2O)_6]^{3+}$, $3d^1$, has only one d-d absorption in pure octahedral symmetry, corresponding to $t_{2g} \rightarrow e_g$. However, in the excited state, the Jahn-Teller effect splits the two eg levels, causing one conformation (e.g. 4 long + 2 short, or 4 short + 2 long bonds) to be favoured.

Depending on the conformation state during the zero point vibrations, one sees a transition to both the eg levels. Since they have different energies, the transition appears as two peaks rather than one. In $[Ti(H_2O)_6]^{3+}$ the bands are at 17.4 and 20.3 kK showing that the magnitude of the Jahn-Teller effect is much less than the value of Δ. Cases where the distortion may be measured (e.g. by X-ray crystallography), are said to show the static Jahn-Teller effect.

Where the stabilisation distortion is similar to the vibrational zero-point energy, no permanent distortion is seen, but there is a distortion of the complex which changes direction rapidly and gives a time average of zero. The Jahn-Teller theorem is not specific to metal complexes, and there is evidence for a Jahn-Teller distortion of the CH_4^+ ion (groundstate $^2T_2$) from photoelectron spectroscopy; however, orbital degeneracy is comparatively rare in compounds not containing partially filled d or *f* shells.

The *second-order Jahn-Teller effect is* the mixing of orbitals resulting from the lower symmetry of a species after distortion. For a linear triatomic molecule the central atom s and p orbitals transform respectively as ag and ir,, but on bending both have a, symmetry, and consequently can mix together—this mixing by distortion is an example of the second-order Jahn-Teller effect—it is usually invoked to explain large distortions and the mixing of levels which were initially of different symmetry.

Transfer spectra

Apart from d-d transitions, most complexes show intense absorption bands resulting from excitation of electrons from orbitals localised mainly on the metal to an orbital localised mainly on the ligands, or vice versa. In the case of WCl_6 where the metal t_{2g} orbitals are unoccupied, transitions from ligand orbitals such as the tin π bonding orbitals are allowed by the dipole moment mechanism, and give rise to intense absorptions (for WCl_6 at 22.4, 26.4 and 29.9 k). In view of the localisation of the orbitals concerned, these bands are referred to as *charge transfer* bands. In general, at least one of the possible charge transfer bands is allowed by the dipole moment selection rules, and consequently the absorptions are very strong, obscuring any d-d transitions nearby.

Their position naturally depends on the relative energies of the metal and ligand orbitals. If we consider ligand to metal transfer (as for WCl_6), the higher the oxidation state of the metal, the lower the energies of the d orbitals. If we consider the bromide complex $[IrBr_6]^{3-}$, the equivalent absorption is seen at 36.8 kK indicating the t_{1u} levels of bromine to be less strongly bound than those in chlorine. These transitions have molar extinction coefficients of about 2 x 10^4 compared with about 50 for the d-d transitions in $[IrCl_6]^{3-}$; the d-d transitions in $IrCl_6^{2-}$ are masked by the t_{1u} $(\pi) \rightarrow t_{2g}$ transfer band.

There is a fair amount of additivity in the positions of these charge transfer bands, and the frequency can be expressed for X-M (ligand to metal) charge transfer as:

$$v_{corr} = [\ X_{opt}(X) - X_{opt}(M)\ ] \ . \ 30kK \qquad (1)$$

where v_{corr} is the frequency corrected for electron repulsion effects, and X_{opt} is the *optical electronegativity*. The correction of v is necessary to avoid anomalies arising from electron repulsion effects, For a high spin d^5 ion, for example, electron transfer will involve spin pairing, while this is not necessary for d^4 ions—the d^5 species wit, thus absorb at a higher frequency even if the t_{1u} - e_g gap is the same. If the frequency can be corrected, then $X_{opt}(M)$ shows a steady rise with the number of d electrons in the ion, and has the form for the dq ion $X_{opt}(M(K)) = \gamma + q\delta$ where γ and δ are the same for a given oxidation state K in a given period, and increase with K. Ligand to metal electron transfer corresponds to simultaneous oxidation of the ligand and reduction of the metal, and the energies of the absorption bands and the values of X_{opt} agree with this: oxidising metal ions such as Mn(VII) show electron transfer in the visible part of the spectrum (e.g. in $KMnO_4$); similarly, the readily oxidised iodide ion has a lower X_{opt} than the fluoride ion.

The treatment of metal to ligand transfer is essentially similar, but in this case the metal should be in a low oxidation state to lower the energy of the transfer—an example is the acetylacetonate $(CH_3\text{-}CO\text{-}CH\text{-}CO\text{-}CH_3)^-$ ion ($acac^-$) which shows metal-ligand (or inverted) transfer for $[Ti(acac)_3]$, but ligand metal transfer for the more oxidising $[Fe(acac)_3]$. Electron transfer is not specifically associated with transition metal compounds: absorption spectra due to a transition from an orbital localised on one atom to one localised on an adjacent atom are found in all branches of chemistry. For example, some of the far ultraviolet absorptions of alkali halides have been ascribed to transfer of an electron from the halide ion to the metal ion. Similarly the yellow colour of $[PbCl_6]^{2-}$ comes from a charge transfer band in the ultraviolet. The transitions involved as electron transfer, or internal transitions, in this case f $\rightarrow$ f and f $\rightarrow$ d transitions.

The electron transfer spectra are little different from those of the transition metals, and optical electronegativities may be derived exactly as for the transition metals. Our main interest is in the transitions involving electrons in the f orbitals. Certain lanthanide and actinide ions, notably Ce(III), Sm(II), Eu(II), Tb(III), Yb(II), U(III) and Np(III) shown strong transitions which are quite sensitive to chemical environment.

These are Laporte allowed 4f $\rightarrow$ 5d (lanthanide) or 5f $\rightarrow$ 6d (actinide) transitions. They are only of sufficiently low energy to be visible if the nf - (n + 1)d gap is quite small, This gap increases with increasing oxidation state and it will be noted that many of the ions quoted above are strongly reducing, such as Sm(II). Most spectroscopic studies have been concerned with f $\rightarrow$ f transitions. These are very insensitive to chemical environment, changing little with change of ligand, very weak (ε_0 ~0.01 - 10), and very narrow (there is no vibrational coupling with the ligands).

Their weakness arises from the fact that they are Laporte forbidden and so little involved in bonding that the means of gaining intensity invoked for d-d transitions (vibronic coupling; overlap with ligands) are no longer important. The f-f transitions of the lanthanides thus represent the extreme 'weak field' case and are very little different 'from the spectra expected of gaseous ions with a spherical environment.

Electron repulsion gives rise to a series of Russell-Saunders terms with separations of the order of

10^4 cm^{-1}. These terms are then split by spin-orbit coupling according to their *J* values. These splittings are smaller, of the order of 2000 cm^{-1}. Finally, any ligand field effects will split the separate *J* levels up according to their M_J values—however, as this effect is typically only 100 cm^{-1} or so, it is very difficult to detect in the spectra.

Ligand field splittings are difficult to detect, but a nephelauxetic effect can be detected, although its magnitude is much smaller than for 3d species, typically 4 percent compared with 10-40 percent for the 3d ions. It presumably arises from the 'central field' effect. These remarks apply equally to the actinide elements, but the 5f orbitals show rather more evidence of involvement in chemical bonding: the nephelauxetic effect is more pronounced, and the bands are broader and about 30 times more intense.

A complication is that the spin-orbit coupling constant $\zeta 5f$ is about double that of the equivalent 4f ion, and the electron repulsion parameters decrease. The spin-orbit splittings are thus greater than or equal to the electron repulsion splittings and the Russell-Saunders scheme may no longer be used, even as an approximation. The strict adherence to the selection rules of these electronic spectra enables one to observe unusual transitions; a magnetic dipole transition has been observed in lanthanide complexes, while an electric quadrupole transition was recently reported for $Cs_2UO_2Cl_4$.

Luminescence is also common for lanthanide complexes, an excitation to the ^3p level of Pr^{3+} could decay directly to the ^3H groundstate, or via a long lived intermediate state such as ^1D). Because of the spin multiplicity difference, some of these excited states can be quite long-lived. This is finding use in laser construction, since glasses containing rare earth ions (particularly neodymium) can be 'optically pumped' easily. Rare earth fluorescence is also for some of the cathodoluminescent phosphors used for colour television. The electronic spectra of compounds of main group elements have been little studied by comparison with those of transition and lanthanide and actinide elements. It is unusual to find electrons in weakly bonding, antibonding, or non-bonding orbitals, and absorptions in the visible part of the spectrum are much less common than for the compounds discussed before. We may discuss the spectra observed under following titles:

1) Molecular spectra
2) Charge transfer spectra
3) Rydberg spectra

Molecular spectra involve transitions between molecular orbitals in covalently bonded molecules. The interpretation of each spectrum depends on the structure and composition of each particular molecule, and it is not possible to generalise as, for example, for d-d transitions or to concentrate on a single central atom.

The spectra of covalently bonded solids may often be best discussed in terms of a band structure. Examples of compounds showing such spectra are the halogen vapours, and NO_2 (brown) (all gaseous), yellow Se_4^{2+}, red Te_4^{2+} and orange S_4N_4. Charge transfer spectra are quite common, and the case of $[PbCl_6]^{2-}$ and its salts has already been mentioned.

The fact that the orbitals involved in the transitions are localised implies that charge transfer spectra are observed in compounds with a fair degree of ionic character. Non-bonding orbitals are rarer in compounds of the main group elements; however, in many compounds of the heavy post transition-metal elements (e.g. the series TI-Pb-Bi) the outermost s-orbital is only weakly involved in bonding (Tl^+, Pb^{2+} and Bi^{3+} all have the configuration

[Xe]$4f^{14}5d^{10}6s^2$). If the s-orbital is occupied, then transitions to the empty p-orbital above it are allowed, and directly analogous to the allowed s $\rightarrow$ p Rydberg transitions of atomic spectra. Such transitions are indeed seen in compounds of Tl(I), Pb(II), Bi(III), and also In(l), Sn(II), Sb(III). They are, however, very different from the transitions observed in the corresponding gaseous ions, occurring at much lower energies, and being quite sensitive to the local environment of the atom.

It seems that the absorption bands of ionic crystals may be due either to charge transfer between oppositely charged ions ($M^+X^- \rightarrow MX$) or to Rydberg excitation of one of the constituent ions (e.g. $Cl^-(3p^6) \rightarrow Cl^-(3p^54s^1)$). These Rydberg like transitions are analogous to the 4f $\rightarrow$ 5d transitions shown by some of the lanthanides.

Optical activity

Many molecules have a symmetry so low that they are not superimposable on their mirror image; the formal symmetry requirement for this is the absence of an improper axis of rotation - a frequently encountered (but rather less general) alternative requirement is the absence of a centre or plane of symmetry. The interest of chiral molecules derives from the fact that separate samples of the two enantiomers will rotate the plane of polarisation of a beam of plane polarised light passing through them in opposite directions.

This behaviour constitutes optical activity, and systems where there are unequal numbers of the two enantiomers of a chiral molecule will show optical activity. Species such as $[Co(en)_3]^+$ could show optical activity was put forward as a proof of their octahedral coordination before X-ray crystallography was developed, Similarly, tetrahedral and square coordination could be distinguished if one or the other would be expected to show optical activity.

The main interest of optical activity is in the spectra of transition metals as measured by the *Cotton* and *Faraday* effects. The rotation produced by a given enantiomer is not constant with wavelength, and, in particular changes rapidly with wavelength near an absorption, passing through zero at the absorption maximum. As an absorption is approached, the absorption of left hand circularly polarised light is not the same as that of right hand circularly polarised light. If we denote the two extinction coefficients by ε_L and ε_R then the plot of ε_L-ε_R shows a maximum at the absorption maximum and is known as a *circular dichroism* (C.D.) curve. The phenomena of change in rotation with wavelength (O.R.D.) and differential absorption (C.D.) are related and are generally described as the Cotton effect. In the two figures we have drawn a positive Cotton effect—the equivalent diagrams for a negative effect can be produced by a rotation of 1800 about the wavelength axis.

The Cotton effect is becoming increasingly studied as interest grows in rates of racemisation (the interconversion of two enantiomers, and the equalisation of their abundance) and the difference in reactivities. The frequent occurrence of chiral molecules in biological systems has led to study of the Cotton effect by bio-inorganic chemists. C.D. data may be used to determine the configuration of ligands about the central atom. The Faraday effect has only recently found wide application. When any compound (not necessarily a chiral species) is placed in a magnetic field with a component parallel to a beam of light passing through the sample, circular doctrinism is observed.

A simple physical explanation may be given. Any electric dipole allowed transition involves a change in angular momentum ΔJ of 1 unit. The magnetic field splits the levels involved in an electronic transition according to their M_J values (the Zeeman effect). The magnetic field gives rise to a slight energy difference between the transitions corresponding to $\Delta M_J = +1$ or -1. However, the transition $\Delta M_J = +1$ corresponds to an absorption of a quantum of right-hand circularly polarised (r.c.p.) light, and $\Delta M_J = -1$ to absorption of a quantum of left hand circularly polarised (l.c.p.) light. Because these two transitions are now separated (by the magnetic field) ε_L and ε_R have different maxima, and their difference gives a magnetic circular *dichroism* (M.C.D.) curve.

Polarisation of absorption spectra

In octahedral symmetry, the electric dipole moment operator which produces the transition between two electronic states has t_{1u} symmetry. If the symmetry of the environment is lowered to D_{4h} the electric dipole moment operator now transforms as the I.R. s $b_{2u} + e_u$; the b_{2u} component corresponds to the component of the incident radiation polarised parallel to the z axis, and the eu component to that polarised in the x-y plane. The transition moments for the two different polarisations will clearly be different (since the dipole moment operators are different) and consequently one might expect a dependence of the absorption on the polarisation of the incident light.

This is indeed found to be the case: in species of low symmetry, the absorption bands are polarised. As the polarisation is measured relative to the molecular axis, it is necessary for the molecular axis to be held in a fixed orientation as in a crystal. Thus the study of the polarisation of absorption bands is confined to the spectra of crystals. Sutton discusses the polarisation of the absorption bands of $[Ni(NH_3)_4(NO_2)_2]$ with D_{4h} symmetry. Strong polarisation effects are particularly common in minerals where transition metals are frequently found in low symmetry environments.

Crystals of tourmaline containing iron absorb light polarised perpendicular to the z axis so strongly that the transmitted light is almost plane polarised parallel to the z axis. The variation in absorption of polarised light can be very useful in the assignment of spectra.

Mixed valence spectra

There exist many chemical compounds containing atoms of the same element in different oxidation states; a good example is Prussian Blue $KFe[Fe(CN)_6]$ containing Fe^{3+} ions, and $[Fe(CN)_6]^{4-}$ complex ions containing iron in the +2 oxidation state. Many of these *mixed valence* complexes show very strong optical absorptions—for example $[Fe(CN)_6]^{4-}$ is pale yellow in aqueous solution, and $[Fe(OH_2)_6]3+$ is almost colourless, yet $KFe[Fe(CN)_6]$ has an intense blue colour. A small amount of a charge transfer state is mixed into the ground state wavefunction of the ligand. Thus, if the two different environments of the same element are A and B, containing the element in oxidation state 2 and 3 respectively, then the wavefunction of the ground state is not $|\psi(A(II)\ (B(III))\ \psi\ (\text{other atoms})|$ as might be expected, but

$$\{\cos\alpha\,[\psi(A(II)\ \psi(B(III))] + \sin\alpha\,[\psi(A(III))\ \psi(B(II))]\}$$
$$\psi\ (\text{other atoms})$$

where α is a measure of the admixture of the charge transfer state into the ground state function. The absorption shown by such compounds to a transition to the corresponding excited state:

$$\{\cos\alpha\,[\psi(A(III)\ \psi(B(II))] + \sin\alpha\,[\psi(A(II))\ \psi(B(III))]\}$$
$$\psi\ (\text{other atoms})$$

Thus Prussian Blue has a wavefunction corresponding to KFe(II) $[Fe(III)(CN)_6]$ with a small amount of KFe(II) $[Fe(III)(CN)_6]$ mixed in—a consequence of this is that neither iron atom has an exactly integral oxidation state —this is not very serious as we are more interested in the mixed valence compound as a whole than in the individual atoms in it. Robin and Day classified mixed valence compounds by the value of α, the mixing coefficient, a depends on (i) the energy difference between the two states (A(II)B(III) and A(III)B(II))—clearly if the state A(III)B(II) has a much higher energy, there will be little advantage gained in mixing with the ground state, and (ii) the extent of the interaction between A and B sites: if the A and B sites are adjacent and overlap strongly or if they are linked by one intermediate ligand forming a fairly covalent bond with each site, the delocalisation and mixing will be much more favourable than if A and B are ionic sites separated by some distance in a crystal lattice. Thus α is larger for sulphides than oxides.

Class I mixed valence compounds contain two very different environments for A and B. $\alpha \approx O$ and there is no interaction. These solids are insulators and show no charge transfer band below 27 kK (although they may show spectra typical of their constituent species). An example is Co_3O_4, a spinel with Co^{2+} high spin ions in a tetrahedral environment and Co^{3+} low spin ions in the octahedral sites.

Class III mixed valence compounds have large a values and extensive delocalisation. They are subdivided into classes IIIA and IIIB. In class IIIA delocalisation is found within a discrete molecular unit (e.g. I_3^-, Bi_5^{3+}). They show mixed valence absorption and it is not possible to identify constituent ions (i.e. I_3^- cannot be described as I^+ + $2I^-$). In Class IIIB the delocalisation is spread throughout the whole crystal lattice, and a band structure description is the most suitable. Class IIIB compounds show metallic conductivity and may show strong magnetic exchange.

Class II compounds fall in between Classes I and III,

α is small but non-zero. They show mixed valence spectra in the visible part of the spectrum, but, if they are not obscured, it is possible to see the spectra of the constituent species, or to detect them by other methods.

The two sites are not too dissimilar, and there is often a bridging ligand (as in $[Fe^{II}Fe^{III}(CN)_{II}]^{6-}$ with a single cyanide bridging ligand). They may show magnetic exchange at low temperatures, and are semiconductors. Prussian Blue is a Class II mixed valence compound.

Magnetic properties

Most molecules and ions have no resultant electronic angular momentum in their groundstate. The most important exceptions to this statement are the compounds of d and f block elements where a d or f shell is partially occupied. Electronic angular momentum produces a magnetic moment, and the study of the magnetic properties may thus provide information about the electronic structure. The magnetic moment will arise from the contributions of the orbital angular momentum and the spin angular momentum. If the groundstate has orbital angular momentum lz and spin angular momentum s, both along the z axis, then the magnetic moment in that direction is

$$\mu z = lz + g.sz \text{ Bohr magnetons.} \tag{2}$$

The factor g arises from a relativistic effect, and is very close to 2. For example, to predict the magnetic properties of a transition metal complex such as $[Fe(NH_3)_6]^{2+}$, we must first determine the groundstate, taking into account electron repulsion, ligand field, and spin-orbit coupling effects. If there is no resultant angular momentum, the complex will be diamagnetic, and the only effect of an applied magnetic field will be to induce a small opposing magnetic field in the sample, giving a small, negative magnetic susceptibility: this is *diamagnetism.* If there is a resultant angular momentum, the groundstate will be degenerate, the number of states corresponding to the number of possible orientations of the angular momentum with respect to an applied magnetic field.

On applying a magnetic field, those orientations with rather than against the field will have lower energy, and the corresponding states will have a greater occupation, giving a net magnetisation of the sample, and a positive susceptibility. This is *paramagnetism.*

Two further points should be borne in mind when considering paramagnetism:

(i) The difference in energy between states aligned with or against the magnetic field is small (for most laboratory magnetic fields, less than 1 cm^{-1}), and there is thus a thermal occupation of several of the states of higher energy. This may be treated by the Maxwell Boltzmann distribution.

(ii) We have assumed that the magnetic moment associated with each metal atom does not interact with the others. If this is not the case the substance is said to show *magnetic exchange.*

A full treatment of paramagnetism therefore involves a good deal of calculation, and goes beyond the scope of this book, the treatment given here being mainly qualitative. As is frequently the case in inorganic chemistry, we may however simplify the treatment in certain cases to give simple relationships which have proved useful sources of chemical information.

Magnetic moments and magnetic susceptibility. The molar magnetic susceptibility X_m is related to the magnetic moment μ by:

$$Xm = \frac{n\mu^2}{3kT} \quad (3)$$

where N =Avogadro's number, k =Boltzmann's constant, T =the absolute temperature, and μ is measured in Bohr magnetons (0.927 10^{-20} erg.gauss^{-1}). As we are normally interested in the paramagnetic susceptibility, it is necessary to correct Xm for the diamagnetism of closed electronic shells of the system (e.g. in diamagnetic ligands).

The orbital magnetic moment in a given direction z was proportional to the orbital angular momentum in that direction, given by the quantum number in. In non-spherical symmetry however, the quantum number in is not always well-defined:

$$
\begin{array}{llll}
 & 12, 0> & & z^2 \\
e_g & /\sqrt{2}\ (12, +2> & + 12, -2>) & x^2-y^2 \\
 & i\sqrt{2}\ (12, +2> & -12, -2>) & xy \\
t_{2g} & /\sqrt{2}\ (12, +1> & +12, -1>) & xz \\
 & i\sqrt{2}\ (12, +1> & -12, -1>) & yz
\end{array} \qquad (4)
$$

For each of these orbitals the value of *m* given by the operator ℓ_z is zero. However, if xz and yz are degenerate, they may mix in a magnetic field to give the eigenfunctions 12, + I > and 12, - 1 > which will have magnetic moments. In a octahedral complex the dxz and dyz orbitals will be degenerate, so we may expect an orbital magnetic moment if the 12, + 1 > eigenfunction is populated to a greater extent than the 12, -1 > eigenfunction (or vice-versa); this will occur in a magnetic field if the t_{2g} orbitals are only partially occupied.

In octahedral complexes the e_g orbitals lie at higher energy than the t_{2g} and the magnetic field is not usually strong enough to mix the dxy and $d_{x^2-y^2}$ orbitals, so the magnetic moment due to the $m = \pm 2$ levels is zero, and is said to be *quenched* by the ligand field. If the complex has symmetry lower than octahedral, the dyz and dxz orbitals may no longer be degenerate, and will no longer be mixed by the magnetic field. The orbital contribution to the magnetic moment is then completely quenched by the low symmetry ligand field. The Jahn-Teller theorem predicts a distortion of any system with an orbital contribution to its magnetic moment, but the magnitude of the distortion is generally too small to make the orbital contribution vanish.

Low symmetry in a complex might be expected to result in anisotropy in the magnetic properties, and this is indeed the case. If we imagine an octahedral complex tetragonally distorted along the z axis, with the t_{2g} subshell incompletely filled (e.g. t_{2g}^5), the dyz and dxz orbitals will still be degenerate, and there will still be an orbital contribution to the magnetic moment, but this is fixed relative to the z axis defined by the distortion.

The contribution to the susceptibility of this orbital magnetic moment will thus depend on the angle of the field to the z axis, and the susceptibility will be anisotropic. In general, magnetic anisotropy is found whenever the symmetry is lower than cubic; however, the effect is much greater when there is a considerable orbital contribution to the magnetic moment, and this contribution is fixed relative to the molecular axes. The magnitude of the magnetic field is usually too small to mix levels which are separated by the ligand field splitting energy Δ, or by very large distortions.

There is however a very small amount of mixing between levels as the field increases (for example mixing of d_{xy} and $d_{x^2-y^2}$) giving rise to a slight paramagnetism. The energy of the upper level is so high that it is not occupied thermally, and the magnitude of the effect is thus dependent only on the amount of mixing into the lower level. The effect is dependent only on the field strength, and is referred to as *Temperature Independent Paramagnetism* (T.I.P.)—it is generally a fairly small effect.

Spin angular momentum

Orbital angular momentum may be quenched by a ligand field, especially if the symmetry is low. A theorem due to Kramers shows that, in a system with an odd number of unpaired electrons, it is impossible to quench the spin angular momentum completely. It is therefore interesting to consider the effects of the spin angular momentum alone, and this gives rise to the 'spin-only' treatment. We assume initially that the orbital contribution is either zero or is completely quenched by the ligand field, and that spin-orbit coupling effects may be ignored (see later). We may now use Hund's rule which tells us that the groundstate will be that with the maximum value of S. The effective magnetic moment meff is now given by

$$\mu_{eff.} = 2\sqrt{S(S+1)} = \sqrt{n(n+2)} \text{ Bohr magnetons} \qquad (5)$$

where n is the number of unpaired electrons. In the presence of a strong ligand field, spin-pairing to give a low-spin configuration may be favoured; this will produce a reduction in n, and a consequent decrease in meff. and X_m. The value of μ can thus be used as an indication of high or low spin configuration. Let us consider an example: for Na_3FeFe_6 m = 5.85 B.M., and for $K_3Fe(CN)_6$ m = 2.25 B.M. Both contain the $3d^5$ Fe^{3+} ion, and so we may deduce that Na_3FeFe_6 is high spin and $K_3Fe(CN)_6$ is low spin.

This is quite in accord with the relative ligand strengths of cyanide and fluoride ions as shown by the spectrochemical series. The high spin ion has one electron in each t_{2g} orbital, and therefore we would expect no orbital contribution. The low spin ion has the configuration t_{2g}^5 and one of the t_{2g} orbitals is only singly occupied: an orbital contribution is thus possible, and the value of m is indeed higher than the 1.73 B.M. predicted by a spin-only calculation.

The spin-only formula is most useful for first row transition metals, and for these compounds the magnetic susceptibility is a useful means of distinguishing high and low spin compounds, as the change in p for a high spin—low spin changeover is generally great enough not to be confused by the presence of a small orbital contribution,

Spin orbit coupling

The interaction of spin and orbital angular momenta, or spin-orbit coupling, was discussed. If this effect is at all important, then the separation of spin and orbital contributions is unlikely to work. For the first (3d) series of transition metals, spin-orbit coupling is much smaller than the ligand field and electron repulsion effects, and we may use, to a first approximation, the 'uncoupled' treatment above.

This simplification does not however hold for the other transition series, and the lanthanide (4f) and actinide (5f) series, as the spin-orbit coupling constant ζ increases steadily with Z, the atomic number. The magnetic properties are determined by the electronic ground state. In spherical symmetry this corresponds to the lowest Russell-Saunders term. As the spin-orbit coupling increases, the Russell-Saunders term is split into levels with different *J* values, where J the total angular momentum is the vector sum of L and S. In such a case the magnetic moment is given by:

$$\mu_{eff} = g\sqrt{J(J+1)}; \quad g = 1 + \frac{J(J+1) - L(L+1) + S(S+1)}{2J(J+1)} \qquad (6)$$

The reader will see that if $L = O$, then $J = S$, and the spin only term results.

For non-spherical symmetry (arising from a ligand field for example), it is necessary to use a different treatment as *J* values are no longer useful. There are two choices: 1) a group theoretical treatment using the double groups (spinors), where each spin-orbit coupled electronic state transforms as an I.R. of the double symmetry group. 2) either apply the spin-orbit coupling as a perturbation to our previously determined states or apply the ligand field as a perturbation to previously determined spin-orbit states (in spherical symmetry). Both methods are complicated, and we give here only the qualitative effects of approach 2 a.

(i) The spin-orbit coupling splits the degeneracy of the ground state if there are spin and orbital angular momenta present. Instead of one state there are now two or more states and spectroscopic transitions to this state will be split, or, since the separate lines are not usually resolved, will be broadened by the spin-orbit coupling.

(ii) The spin and orbital angular momenta can either be coupled parallel to each other, increasing the total μ, or anti-parallel, decreasing μ. If they are antiparallel in the lowest spin orbit coupled state, then μ will be reduced below the spin only value, in some cases to zero. If the temperature is raised, then the thermal population of the higher spin-orbit coupled states increases, until, in the limit, the total effect of the spin-orbit coupling is zero, since the excited states with spins and orbital angular momenta coupled parallel will be occupied to the same extent as the lower energy states with antiparallel coupling. We would therefore expect an increase in μ on raising the temperature. If the spin-orbit coupling is large then the value of $\frac{kT}{|\lambda|}$[3] is small and the value of μ changes with the temperature, while if $|\lambda|$ is small the 'plateau' region B is reached at quite low temperatures and the magnetic susceptibility changes little with temperature. For the $3d^n$ transition metals, the spin-orbit coupling is small, and at all but the lowest temperatures the magnetic behaviour corresponds to region B. For second and third row transition metals, X is much larger and the magnetic behaviour is better described by region A. Thus simple 'spin only' treatments of complexes of the heavy metals are very unreliable.

(iii) Spin-orbit coupling, by mixing the spin and orbital parts of a wavefunction increases the intensity of formally 'spin forbidden' electronic transitions. The two states involved in a transition do not have well-defined values of S, and consequently ΔS is not well defined.

(iv) Just as electron repulsion parameters are affected by the environment of the ion (the nephelauxetic effect) the spin-orbit coupling parameter is sensitive to chemical environment, and is reduced below the free ion value.

Calculation of paramagnetic properties

The method of calculating magnetic properties have many ways: first it is necessary to calculate the magnetic moments associated with each of the possible states of the system, second to calculate the thermal occupation of these states (using the Boltzmann distribution) and sum the product of the population and the magnetic moment of a given state to give the bulk property.

The second step is simple classical physics, but the first step involves using ligand field theory. One proceeds by a series of perturbations, applied in order of decreasing magnitude. Thus for a weak field example we could start with pure atomic d functions, apply the electronic repulsion interaction to obtain the wavefunction of the lowest Russell-Saunders state. The next step is to apply the ligand field - there are two possibilities here: 1) we may mix in some ligand orbital character - this is normally only done for a strong field case; 2) we may simply multiply the wavefunction by an orbital reduction factor k, which is less than unity, to allow for any covalence - this is a simpler procedure, although its physical meaning is clearly rather dubious.

We now have a wavefunction, or set of degenerate wavefunctions for the ground state of the metal atom in the complex; we still have to apply the spin orbit coupling and magnetic field perturbations. It is necessary here to decide whether any excited states will be mixed into the groundstate by these two remaining perturbations. If the complex is pure octahedral, then the excited states involving the e_g orbitals will be at least 10^4 cm^{-1} above the groundstate for a first row transition metal; the spin orbit coupling will typically be 300 cm^{-1}, and the effect of the magnetic fields normally used even smaller (~1cm^{-1}); it is thus unlikely that they will mix these excited states into to groundstate, and they may be ignored. If there is a slight distortion, (perhaps as a result of the Jahn-Teller effect) there may be low lying excited levels (say at a few hundred cm^{-1}) which may be mixed in, or even partially occupied thermally, and these may not be neglected (kT at 300 °K ~200 cm^{-1}).

The spin orbit coupling perturbation must then be applied to first and second order, considering first the lifting of degeneracy of the groundstate and then any mixing in of excited states. Finally, we should consider the effects of the magnetic field—this normally need only be done to first order, splitting the groundstate into terms with different alignments to the magnetic field. If the second order term is considered, the mixing in of states which gives rise to temperature independent paramagnetism appears.

The relative magnitudes of the effect discussed are not the same for all d_n and f_n configurations. In Table 1 we give examples of how the magnitudes of these effects vary. Thus for lanthanides we see that electron repulsion > spin orbit coupling> ligand field effects, the ligand field effects being so small as to be almost swamped at room temperature (kT = 200 cm^{-1}) - we may ignore ligand field effects, and use Eq. (6) relating meff, to J. For 4d and 5d elements, the ligand field effects are stronger than electron repulsion effects, suggesting a strong field approach to be suitable,

Table 4

Relative magnitudes of effects

Metal	Electron repulsion	Ligand field	Spin-orbit coupling meter *D*	Spin-pairing energy para
$3d^n$	12000	15000	400	6000
$4d^n$	8000	22000	1000	4000
$5d^n$	5000	25000	2500	3000
$4f^n$	15000	50	1500	6000
$5f^n$	10000	100	2500	3000

Magnetic exchange phenomena arise when the magnetic moments of individual complex molecules interact and couple—it is then no longer possible to consider each molecule as being separate and unaffected by the others. Magnetic exchange phenomena thus affect the bulk properties of the solid.

The magnetic moments of the individual complex species are all coupled together within a certain region (or domain); if the solid is very finely ground, the size of the individual particles can become so small that there are not enough spins present to show the normal magnetic exchange interaction, and the particles show normal paramagnetic behaviour—this phenomenon is known as *superparamagnetism.* Ferromagnetic species have the magnetic moments coupled parallel to each other, giving a large total magnetic moment, and a large magnetic field inside the solid. This leads to a large positive susceptibility. Ferromagnetism is generally found in metals and alloys, and is best known in iron metal.

Ferromagnetic materials become paramagnetic above a certain temperature (the Curie point) above which the thermal energy is too great to maintain the alignment of spins. Antiferromagnetic species have their spins coupled antiparallel to each other so that each magnetic moment is balanced by an equal and opposite moment. This leads to a zero magnetic susceptibility. Again the coupling breaks down above a certain temperature (the Neel temperature) and normal paramagnetic behaviour is shown.

A decrease in p with decreasing temperature is sometimes taken as an indication of anti-ferromagnetism, but can often be ascribed to spin-orbit coupling, effects reducing υ. It should be noted that antiferromagnetic 'pairing' involves no actual mixing of orbitals (as for formation of a molecular orbital) but may be regarded as a 'through space' interaction in which the state with moments opposed is favoured. Antiferromagnetism is common in compounds (such as FeF_2, Neel temperature 77 °K, K_2IrCl_6 3.08 °K). The mechanism of exchange is not well understood, but it may be remarked that anti-ferromagnetism is common in closely packed structures with electronegative atoms (such as F, Cl, O) present.

Ferrimagnetism is the simultaneous manifestation of ferro- and anti-ferromagnetism; in a solid with two different sites occupied by paramagnetic species, the paramagnets in one site may couple ferromagnetically with themselves and antiferromagnetically with the other paramagnetic site. If there is an imbalance in the population of the two sites, this leads to a net alignment of the spins and a large magnetic field in the solid and a large susceptibility. Solids showing such ferrimagnetism are referred to as ferrites. An example is the spinel structure which has two possible metal cation sites within a cubic close packed array of oxygen atoms.

AB_2O_4 has, for each 16 oxygen atoms, 8 octahedral sites and 16 tetrahedral sites; in Fe_3O_4 the tetrahedral and half the octahedral sites are occupied by Fe^{3+} ions, and Fe^{2+} ions occupy the other eight sites. The magnetic moments of the two Fe_{3+} sites cancel out, leaving the coupled moments of the Fe^{2+} ions— giving the large magnetic susceptibility of magnetite Fe_3O_4 Many artificial ferrites have now been prepared, some of which have extremely useful magnetic properties. A complex is high spin or low spin according to the relative magnitude of electron repulsion and ligand field effects, and we may imagine circumstances in which these two effects might be roughly equal.

In such cases the complex may be found in both high and low spin states, depending on the experimental conditions, and is said to exhibit spin crossover behaviour. Thus the octahedral $3d^6$ complex $[Fe(phen)_2(NCS)_2]$ has a magnetic moment corresponding to high spin Fe^{2+} at room temperature, and is diamagnetic (low spin Fe^{2+}) at low temperatures. The change from high spin to low spin behaviour (or vice-versa) can be produced by changing temperature or pressure, and presumably results from a change in the metal-ligand bond lengths, and the possible occupation of electronically excited states. Spin crossover occurs in various ways.

For the high spin—low spin crossover of Fe^{2+} mentioned above, there appear to be two phases (one high spin, the other low spin) coexisting near the crossover temperature, and distinct Mossbauer spectra are seen for the different sits.

Chapter 2

Molecular Orbital Theory

One approach to simplifying the mathematics is to begin with the functions that describe the individual atomic orbitals and then combine them. The result is a mathematical description and a physical picture of a new orbital—a molecular orbital. A molecular orbital is the space in which an electron with a specific energy is most likely to be found in the vicinity of two or more nuclei that are bonded together.

The electron can be closer to one nucleus at one instant and closer to the other nucleus at another instant. The difference between bonds in the molecular orbital and valence bond theories is a little bit like the difference between scrambled eggs and two fried eggs that have run together at the edges in the frying pan. Molecular structure and geometry are more easily pictured in terms of the VSEPR and valence bond theories than in terms of molecular orbitals.

However, the valence bond theory does not clarify the relative energy levels of the orbitals involved in the bonding—something that can be done by using molecular orbitals. An understanding of energy levels is important—for one reason, because we derive much of our information about molecules from spectroscope, to interpret spectra, a knowledge of the number of possible excited states for electrons and of the energies of the transitions to and from these states is needed. One of the great successes of molecular orbital theory is the prediction and explanation of electronic spectra. Molecular orbital theory also permits a clearer understandirg of multiple bonds, molecules for which resonance must be used to write Lewis structures, and molecules with unpaired electrons.

The molecular orbital approach, like the valence bond approach is based on quantum theory and the use of the Schrodinger equation, The energies of electrons in bonds, like those in isolated atoms, are quantized, For molecules, not only the interactions of electrons with each other and with nuclei must be accounted for, but also the interactions of nuclei with each other. As for the atomic orbitals of many-electron atoms, approximations are used to reduce the complexity of the mathematical treatment of molecular orbitals. Energy-level diagrams can be drawn for molecular orbitals.

Then, as in the case of atoms, electron configurations can be determined by distributing electrons among the energy levels. From the configuration of the electrons in the molecular orbitals, predictions can be made about whether or not a bond between two specific atoms will form and whether or not a molecule will be paramagnetic—information that is not easily available from valence bond theory.

For example, molecular orbital theory explains why Be_2 molecules have only a brief existence and it can account for the magnetic properties of O_2 molecules, neither of which can be done easily by valence bond theory. In reading the following sections, keep in mind that we are describing a mathematical process for representing bonding as it is observed. In this chapter we shall apply molecular orbital theory to a wide variety of compounds in order to show that it may be used as a quite general means of description.

Simple polyatomic species

Here we will examine the structures of some compound of elements having only s and p type atomic orbitals in their valance shell.

ML_2 molecules

In discussing MgF_2 we considered only the linear structure known from experiment. We may now consider the M.O.s of a bent ML_2 molecule, assuming that the two M- L bonds are of equal length. This molecule has $C2_v$, symmetry, with a twofold symmetry axis along the z axis. The orbital energies will determine which of the s or p_z is more involved in bonding. In Fig. 1, a molecular orbital diagram for a ML_2 molecule is shown.

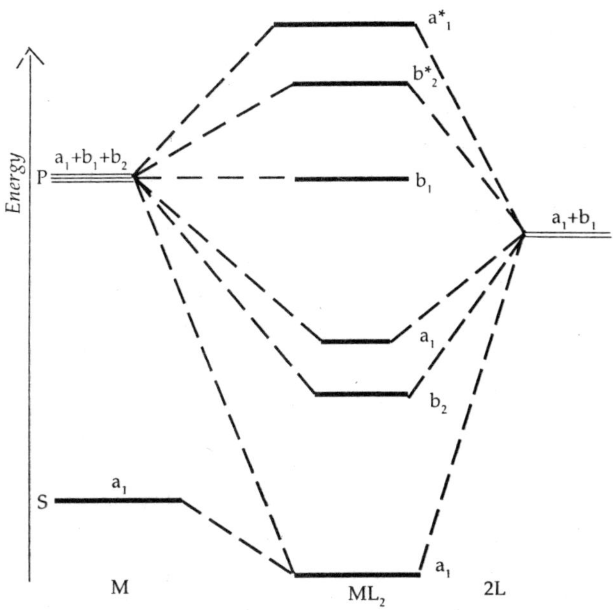

Fig. 1: The molecular orbital diagram for a bent ML_2 molecule

The b_1, orbital is non-bonding, and the a_1 (bonding) orbital will be raised slightly above the b_2 (bonding) orbital by interaction with the a_1 (s) orbital. The order of the a_1* and b_2* orbitals is uncertain. We may, now consider what happens if we increase the L-M-L angle until the molecule is linear along the y axis. The b_2 orbital becomes a σ_u bonding orbital, and the other two p orbitals of M become non-bonding π_u orbitals. The a_1 (s) orbital correlates with the a bonding M.O. As far as the orbital energies are concerned, the p_x orbitals is always non-bonding, and its energy will not change greatly on becoming linear, but the p_z orbital changes from bonding to non-bonding in the linear molecule, and its energy therefore rises.

The b_2 (p_y) orbital falls in energy in the linear molecule, since the overlap with the ligands will be greater, and the ligand b_2 orbital is more L-L antibonding in the bent molecule. The a_1(s) orbital will not change greatly in energy on distorting the molecule. It will be seen that a molecule with four or less electrons in the valence shell will prefer linear geometry, but that further electrons will enter the π_u/a_1 orbital which is strongly stabilised by bending.

Experimental data support this conclusion. BeH_2 with 4 valence electrons is linear, but BH_2 with 5 valence electrons is bent; excitation of one electron to give the excited state of BH_2 $(a_1)_2$ $(b_2\ )_2$ $(b_1)_1$ reduces the stabilisation of the bending, and the excited state is linear or nearly linear. Singlet CH_2 with two electrons in the $a_{1/}$ π_u orbital is strongly bent, but the triplet form with one electron in the $a_{1/}\pi_u$ orbital and the other in the b_1 orbital is less so. NH_2 and H_2O with respectively 7 and 8 electrons are bent. Further experimental support for this description comes from the unpaired electron spin in species such as PH_2.

The electron spin couples with any nuclear spin present in the molecule, and, if the magnitude of the coupling with a given nucleus can be measured, one may estimate to what extent the impaired electron is localised on that atom. For PH_2, where the impaired electron is in the b_1 orbital, the coupling with the phosphorus nuclear spin is much stronger than with the proton spin, confirming it as a non-bonding orbital localised on the central atom. Photoelectron spectra have now shown that the s orbitals of the central atom are rather less important than originally thought and the description given above has taken account of this.

Consideration of M-L multiple bonding does not fundamentally change the approach. In most cases (e.g. MgF_2) the L π orbitals lie at lower energy than the Mπ orbitals, and, for linear geometry, the π non-bonding entirely localised on the ligands. These orbitals change relatively little on bending, and the structure adopted depends mainly on the occupation or non-occupation of the $\pi_u{}^*/a_1$ orbital, as for the σ–bonding only case.

The π_u (bonding) and π_g (non-bonding) M.O. lie below the $\pi_u{}^*$ and will be filled first. On progressively filling these orbitals, we expect linear geometry until the configuration $(\sigma_g)^2$ $(\sigma_u)^2$ $(\pi_u)^4$, $(\pi_g)^4$ $(\pi_g)^4$ $(\pi_u{}^*)^2$ is reached, at which point bending will be favoured. Both NO_2 and CO_2^- have this configuration and are bent, whilst CO_2, NO_2^+ and MgF_2 have one less electron and are linear. For simple triatomics where π bonding is possible, up to 12 electrons may be accommodated in the M.O.s described above without bending; it is usual to count as well the two pairs of electrons in the non-bonding 1s shells of ligands such as F, O and N, and the critical number of valence electrons in these systems is thus 16. 17 and 18 electron species will be bent, as will 19 and 20 electron species since the $\pi_u{}^*/b_1$ orbital above the $\pi_u{}^*/a_1$ orbital is almost unaffected by bending (it is actually slightly stabilised). 21 and 22 electron systems involve filling the

σ^* M.O.s which are strongly destabilised by bending, and are consequently linear 24 electron systems have all the antibonding orbitals filled and are thus unstable. Table 1 shows the measured bond angles for a large number of ML_2 molecules. The geometric predictions of the Walsh diagram are verified.

Table 1

Structures of simple triatomic molecules

Molecule	No. of valance electrons	Bond angle
CO_2	16	180
NO_2^+	16	180
CO_2^-	17	127 ± 8
NO_2	17	134
NO_2^-	18	115
CF_2	18	105
SiF_2	18	101
GeF_2	18	94 ± 4
O_3	18	116.8
SO_2	18	119.5
ClO_2	19	117.5
F_2O	20	103.3
Cl_2O	20	110.9
ClF_2^+ —bent	20	
ClF_2^- —linear	21	
NNO	16	180
SSO	18	118

We may also note that the N-O bond length in the series $NO_2^+ : NO_2 : NO_2$ increases from 1. 10 4→1.19 → 1.24 Å implying that the new orbital being filled has antibonding character, as predicted by the diagram. Herzberg discusses the electronic spectra of many simple compounds, and relates them to Walsh diagrams. Walsh also pointed out that the loosest hound (and presumably most reactive) electrons in NO_2 and NO_2^- are in the π_u^*/

a_1, orbital of the molecule, and that, as this M.O. is localised mainly on the nitrogen atom, reactions of these species will involve the nitrogen atom rather than the oxygen atoms. This is generally observed, and is also found for the sulphur atom in SO_2. The effect of orbital energy variation on geometry is not easy to predict. One might reasonably assume that, if the s orbital of M is unwilling to take part in bonding, the contribution of the p_z orbital will be more important, and the bending will be more favourable.

This will produce a variation in bond angle, as is seen in Table 1, but will not cause a change from linear to bent geometry (or vice-versa) for a given number of valence electrons. This is another example of the structure being determined by the number of valence electrons, irrespective of the variation in orbital energy (of. BH_4; CH_4; NH_4^+). Electron counting also works for molecules which have lower symmetry than that assumed above.

Thus $ICBr^-$, NNO, or SSO have geometry and spectra similar to those predicted for molecules which are symmetric about the central atomic appears that the lower symmetry perturbs the molecular orbital levels (and in theory allows more mixing between atomic orbitals) but does not change the pattern of the levels sufficiently to nullify the predictions of the Walsh diagram.

ML_3 molecules

We consider two possible structures: planar (Symmetry D_{3h}), and pyramidal (symmetry C_{3v}), and consider only the effects of σ bond formation with the ligands L. For D_{3h} symmetry. The orbitals concerned transform as the the following I.R.s:

M s	a_1''
Mp_x, p_y	e'
Mp_z	a_2''
3Lσ	$a_1' + e'$

The M p_z orbital is thus non-bonding, and the interaction will be between the M s, M p_z and M p_y and the ligand a orbitals. If we assume the ligand σ orbitals to be intermediate in energy between the N_4 s and M p orbitals, the M.O. diagram is as shown in Fig. 2a, and there are three stable bonding M.O.s, and one non-bonding M.O.

The structure will be quite stable for up to six valence electrons—it is in fact the stable form for BF_3 with six valence electrons. The pyramidal structure may be obtained from the planar by pushing the central M atom along the threefold axis, out of the plane of the ligands. In this symmetry (C_{3v}) the orbitals transform as:

M s	a_1
Mp_x, p_y	e
Mp_z	a_1
3Lσ	a_1+ e

The p_z orbital is obviously no longer non-bonding and consideration of Fig 2a shows that it may interact with the a_1* orbital, causing one orbital to rise in energy, and the other to fall (as was the case when the σu interaction was introduced for MgF_2). The M.O. diagram is given in Fig 2b the pyramidal structure will thus he more stable than the planar if there are eight electrons involved in the bonding, and the second a, orbital is occupied. NF_3 and NH_3 do indeed have this structure.

We may also note that this highest occupied orbital (a) will have a fair amount of Mp_z, and σ orbital

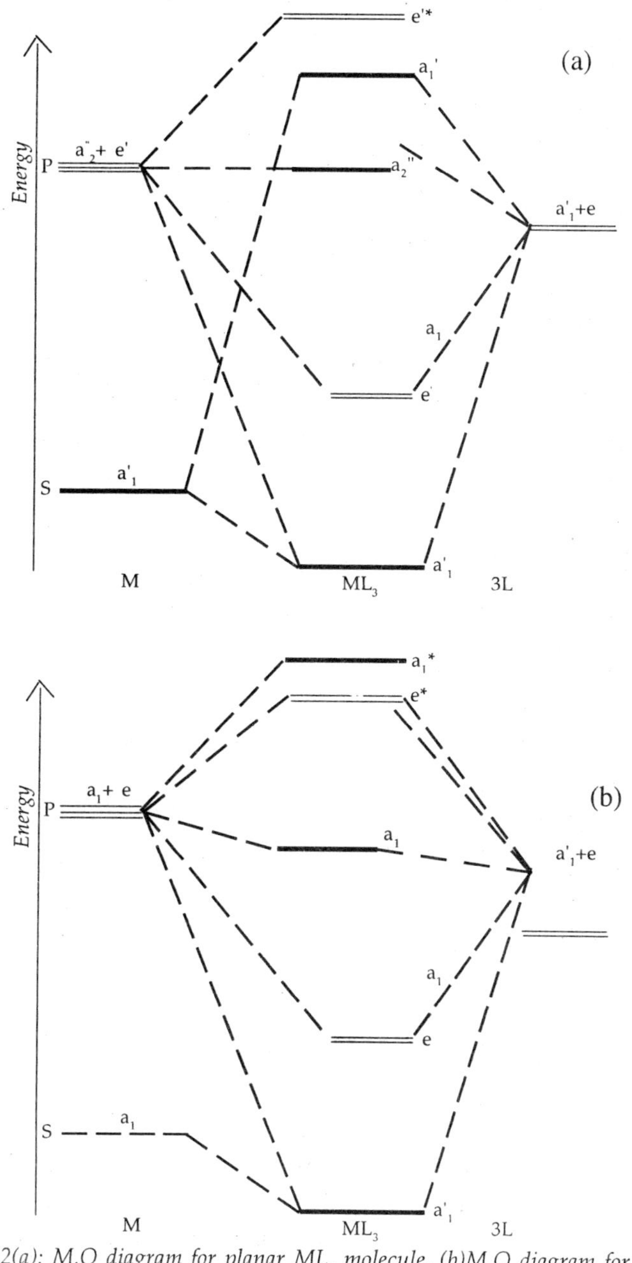

Fig. 2(a): M.O diagram for planar ML_3 molecule, (b)M.O diagram for pyramidal ML_3 molecule

character and will be directed away from the L_3 plane: many pyramidal ML_3 compounds are good electron pair donors (or Lewis bases) as a result of this orbital being occupied. If the M s orbital is virtually non-bonding, either as a result of a low overlap integral with the ligands, or because it is tightly bound, the bonding will involve the M p orbitals only, and the bond angles will be close to 900. This is observed for the series PH_3, A SH_3 and SbH_3, where the bond angles are much closer to 900 than for NH_3.

The planar BF_3 molecule has a low lying, empty, non-bonding orbital (a_2'') localised on the boron atom. This orbital can accept a pair of electrons, functioning as a Lewis acid, however, in accepting a pair of electrons from a Lewis base (such as ammonia), it increases the number of electrons in the valence shell, and distortion to a pyramidal structure is now favoured, as the energy of the acceptor (a_2''/p_z) orbital will be lowered by the distortion. This is indeed observed, and BF_4, the fluoride ion adduct of BF_3 is tetrahedral. In passing we may note that the planar structure involves only M s, M p_x and M p_y orbitals of the central atom, and is thus close to sp^2 hybridisation.

The pyramidal structure uses all s and p orbitals of M, and corresponds to sp^3 hybridisation. The comparison of the M.O. diagrams for different structures and the construction of Walsh diagrams enable us to predict the more stable structure for a compound. In the two cases discussed it was seen that the occupation or non-occupation of one M.O. was critical in determining the structure adopted.

Some more complicated examples are discussed in a review by Burdett. For compounds Where rr bonding is relatively unimportant, it is often possible to localise bonds and predict the stereochemistry about a particular atom in quite complicated molecules. Tertiary phosphines and amines (R_3P and R_3N) are pyramidal even if the three organic substituents are different, each carbon atom of the C—N (or C—P) bonds may be assumed to have an identical localised orbital available for bonding.

Molecular orbital energies

The energy of molecular orbitals is not easy to determine: ultraviolet and visible spectra determine only the energy differences between orbitals and not their absolute energies. The recent development of photoelectron spectroscopy has enable the vertical ionisation energies of specific molecular orbitals to be measured, and, consequently, an estimate of their binding energy to be made. In all cases the number of bands seen corresponds with the pattern of orbitals predicted by simple molecular orbital theory. The central atom Is orbital is always too tightly bound to have any involvement with chemical bonding.

The orbital formed from the hydrogen atomic orbitals and the 2s orbital can be clearly identified in CH_4 ($2a_1$) and the corresponding band in the other compounds rises monotonically in energy to the value for the pure 2s orbital in neon. This suggests that this orbital remains fairly close to a pure 2s orbital, although some mixing with a, type p orbitals is possible in molecules of lower symmetry (NH_3, H_2O, HF).

The weaker bound electrons must therefore be either non-bonding p orbitals, or p orbitals bonding with the ligand σ orbitals. It is instructive to take the mean p orbital energies and compare them with the s orbital energies. As one crosses the series, the ratio Es:Ep become much greater, P implying that the 2s orbital rises in binding energy much faster than the 2p. For HF the energy difference is 22 eV (~2000 kJ/mole), and, as the bond energy is only 556 kJ/ mole, extensive s orbital participation in the bonding seems improbable, For NH_3 the s-p gap is much smaller (almost the same as the total NH bonding energy), and interaction with the s orbital seems more probable. The H-N-H angle is much greater than the 900 expected for p orbital bonding only; for H_2O the s type orbital $2a_1$, is more tightly bound, and, in accord with this, the bond angle diminishes, implying less s character.

Transition metal compounds

All transition metals have a partially filled d sub shell in the free atomic state, and chemists were inclined to think that this required a separate theoretical treatment of the bonding in their compounds. One of the purposes of this section is to show that the principles of this treatment are identical to those already established.

Reference to the sketches of the *d* orbitals, as they do not have any particularly well established directional properties, with four lobes of electron density pointing in different directions. We are not surprised then to find that elements where the d-orbitals play a strong part in the bonding generally have several atoms bonded to them, or, to put it more elegantly, such elements tend to show high coordination number (C.N.), rarely less than 4, most frequently 6, and sometimes as high as 9.

We may note in passing that elements with partially filled f shells often show even higher coordination numbers, although this does not appear to be directly related to the presence of occupied f orbitals. This tendency to high C.N. is shown not merely in the solid, but also in solution, where central atoms with d-orbitals involved in bonding will keep their ligands (i.e. surrounding atoms or groups) quite strongly bond. This was established by the work of the Swiss chemist, Werner, at the end of the nineteenth century, and the high C.N. species were described as complexes; for many years the high C.N. of the central atom in complexes posed considerable theoretical problems for chemists who had based their bonding theories on elements such as carbon and nitrogen usually showing low C.N The fact that d-orbitals do not 'point' strongly in one particular direction results in overlap integrals frequently being much smaller than for p orbitals. Many complexes show quite high symmetry, and even those which do not may be discussed in terms of a high microsymmetry. For example, the complex $[Co(NH_3)_6]^{3+}$ has the ammonia groups octahedrally disposed about the cobalt atom (symmetry O_h).

Substitution of one ammonia by a chloride ion gives the complex $[Co(NH_3)_5$ where the disposition of the ligands ($5NH_3$, 1Cl) is still octahedral, but the symmetry is lower (C_{4v}). Although the energy levels are slightly perturbed, the electronic structure can still be discussed usefully in terms of microsymmetry O_h. A complex is conventionally described as octa-hedrally coordinated if there are six ligands at the apices of an octahedron centred on the central atom. This does not imply that the symmetry is octahedral.

Thus $[Co(NH_3)_5$ CH_2, has octahedral coordination, but lower symmetry. We may frequently be able to discuss an octrahedrally coordinated species in terms of an octahedral microsymmetry, but this is another assumption. We have already seen an example of this with NO_2 and SO_2, which may be discussed in terms of microsymmetry C_{2V}.

Electronic structure of octahedral complexes

Tungsten hexamethyl is a well characterised, if rather reactive octahedral complex of tungsten with symmetry O_h. The methyl ligands have no orbitals suitable for π bonding and we will consider that they have available one sigma orbital each for bonding with tungsten. In view of the closeness of the CH bond angles in methyl groups to the tetrahedral angle of 109.5° this orbital will probably be close to a sp^3 hybrid, but it is really only its σ bond symmetry which interests us here. We place the six groups on the x, y and z axes. We now need to consider the metal A.O.s.

The only A.O.s available for bonding are the 5d, the 6s and the 6p, although these last may have too high an energy to take much part in bonding. All three sets of orbitals are sufficiently diffuse to overlap well with the ligand groups. Although the 6s and 6p each transform as one I.R. of the symmetry group (a_{1g} and t_{1u}), the 5d orbitals transform as t_{2g} and e_g.

The e_g set consists of the d_z2 and $d_x2\text{-}y2$ orbitals —it is thus not surprising that the d orbitals transform as two different I.R.s. The mean energy of the 5d orbitals is below that of the 6s in tungsten, and thus we arrive at the level of A.O.s shown in Fig. 3. We have put the level of the CH_3 σ orbitals slightly below that of the tungsten 5d, in accord with general views about their energy. The construction of the molecular orbital diagram is now quite straightforward: the t_{2g} orbitals will be non-bonding, and thus their energy will change very little.

The two sets of e_g orbitals will give one set of bonding and one of antibonding molecular orbitals. The a_{1g} orbitals will also give a bonding and an antibonding set, but the greater energy separation will give a weaker bonding interaction, and thus the a_{1g} bonding orbital will lie above the e_g bonding orbital. A similar consideration will apply to the t_{1u} sets, which will interact even more feebly. We thus arrive at the molecular orbitals for $W(CH_3)_6$ (Fig. 3).

Tungsten has 6 valence electrons, and the methyl groups will have one electron in each of their σ orbitals we thus have 12 electrons to fill the molecular orbitals, which just fills the six bonding molecular orbitals eg, a_{1g},

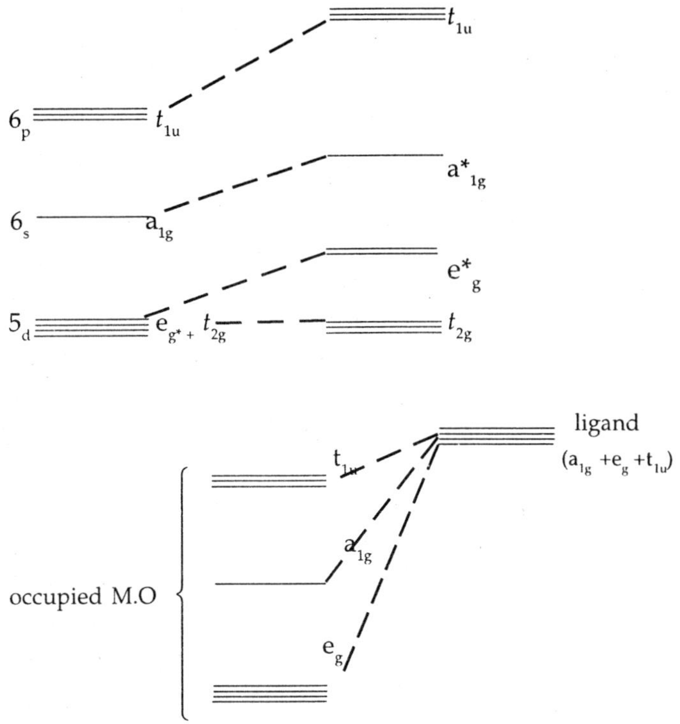

Fig. 3: W W($(CH_3)_6$ $6CH_3$

We may classify the orbitals in the following order:

e_g / a_{1g} } bonding	e^*_g / $a_1{}^*{}_g$ } antibonding
t_{1u} weakly bonding	
t_{2g} nonbonding	t_{1u} high energy, weakly bonding

t_{1u}. We would thus predict that the compound would be stable as indeed it is (to dissociation—it is reactive simply because tungsten and methyl groups can form stronger bonds with other groups, for reasons such as improved overlap). Let us now consider the nature of the first four sets of molecular orbitals. The e_g orbitals are composed of We_g (5 d) orbitals and CH_3 e_g orbitals. Because the energies of these two sets are close, the M.O. will probably have roughly similar amounts of metal and ligand character i.e. the bonding is quite covalent.

The a_{1g} orbital will be mainly methyl σ orbital (because of the higher tungsten 6s energy), and the tlu probably almost all methyl σ orbital in character. The unoccupied t_{2g} orbitals will be almost unchanged tungsten 5d orbitals. The 6 occupied orbitals will have rather more methyl than metal character, and since each provides 6 electrons, there will be a slight loss of charge to the ligands.

The photoelectron spectrum of this species has been measured and is quite in accord with this, also suggesting that the 6p orbitals play little part in the bonding i.e. that the t_{1u} bonding orbital is almost pure methyl σ in character. The construction of the M.O. diagram has presented no special problems, and because of the high symmetry, has even been slightly easier than (for example) for NO_2. Note that the eg and alg bonding orbitals will be highly delocalised over the metal and six ligands, and that it is not possible to draw suitable localised bonds. Now let us turn to a complex which shows some of the features peculiar to transition metal compounds.

Electronic structure of $[Fe(NH_3)]_6^{2+}$

The ferrous ammonium ion is another octahedrally coordinated complex, and although the symmetry is not quite perfectly octahedral we will assume an O_h, microsymmetry. NH_3 has the same σ bonding orbital as CH_3, but it is now doubly occupied. It is normal to place the NH_3 σ orbitals well below those of the metal, from energy level and other considerations. The molecular orbital diagram will not be substantially different from that for WMe_6, except that the lower energy of the ligand a orbitals will imply that the cr bonding orbitals $a_{1g} + e_g + t_{1u}$ have less metal and more ligand character.

The 3d orbitals of Fe^{2+} are much less diffuse than the W5d, and the overlap with the ligands is less—this leads to a smaller bonding effect, and the e_g^* is raised above the t_{2g} level much less than in WMe_6. In WMe_6 there were 12 valence electrons; in $[Fe(NH_3)_6$ there are 12 in each ligand σ orbital and 6 on the central ion (Fe^{2+}—$3d^6$). We will therefore need to fill more molecular orbitals, and 6 electrons must be placed in the t_{2g} levels, or higher.

The first three may be placed in the t_{2g} level with a clear conscience, but with the fourth electron we must realise that 1) to place it in the t_{2g} orbitals will involve pairing it with one of the three already there (Hund's maximum multiplicity rule requires that the first three have parallel spins), and this will involve a loss of spin pairing energy, and 2) the weak bonding interaction means that the e_g orbitals are only slightly higher than the t_{2g}.

The lowest energy is obtained if the electrons are arranged to minimise the pairing of electron spins—this is achieved if two electrons are placed in the e_g^* levels. The energy difference between the two levels is generally referred to as the ligand field splitting energy and is denoted by Δ or 10Dq. Its value is typically around 15000 cm^{-1} (= 15 kiloKayser (kK) or 2eV, or 200kJ/mole). The term 'ligand field' derives from the idea that the d orbitals are split in energy by a potential field due to the ligands.

The energy 15000 cm^{-1} falls in the visible region of the spectrum, and the visible and near ultraviolet spectra of transition metal complexes have been of incalculable importance in establishing their structure, rather than break this purely theoretical approach we discuss them separately.

If we now move to the next element in the first transition series (cobalt) has seven electrons to be distributed in the e_g* and t_{2g} levels. The ligand $[Co(NH_3)_6]^{2+}$ field stabilisation energy Δ is little changed, and the spin pairing energy is still sufficiently high for the population of the higher energy e_g* to be favoured. However, this complex may be oxidised readily to give $[Co(NH_3)_6]^{3+}$ with one less electron, the e_g*—t_{2g} gap Δ is now much larger (22.9 kK, compared with 10.lkK for $[Co(NH_3)_6]^{2+}$, and the complete occupation of the t2g subset is favoured over partial occupation of both e_g* and t_{2g}, since the gain in spin pairing energy for the occupation of the e_g* is less than the promotion energy, Δ, that is required.

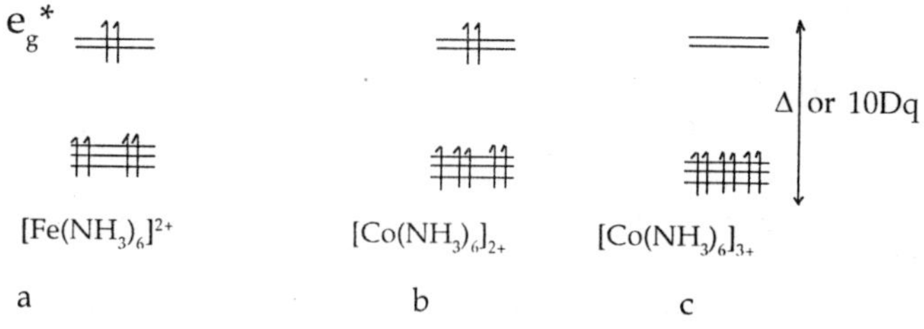

Fig. 4:

Figures 4a and c represent two different ways of arranging 6 electrons in the system. Since both systems have 6 electrons in the t_{1g} and e_g* orbitals which are predominantly d orbital in character, they are referred to as d^6 ions ($[Co(NH_3)_6]^{2+}$ is a d^7 ion).

The arrangement in Fig.4a where the electrons are arranged to give the maximum number of parallel spins (and lowest spin pairing energy) is described as high spin that of Fig. 4c is a low spin arrangement.

A low spin or high spin arrangement is chosen according to whether the ligand field stabilisation energy A is greater or less than the energy required to pair the spins. The reader should be able to show for himself that the ions d_1, d_2, d_3, d_8, and d_9 do not have available to the two alternatives of high spin or low spin configuration in octahedral complexes. So far, we have only considered ligands with one sigma bonding orbital had we considered chlorine as a ligand we would have considered one 3p orbital as having π symmetry, while the other two would have σ symmetry about the metal chlorine bond axis; any bonding involving these orbitals is metal-ligand π bonding.

The nature of interaction depends on the energy of the ligand π orbitals; for chlorine they will be strongly bound, while for ligand groups such as CO there are low lying antibonding orbitals of π symmetry. We will consider the two cases separately. The effect of the ligand π orbitals at low energy is to raise the energy of the predominantly metal t_{2g} and t_{lu} levels, now formally π antibonding; the $t_{lu}(p)$ orbitals were already weakly σ antibonding, and the π interaction would be expected to be even weaker.

The π bonding t_{lu} have been put only just below the nonbonding tlg and t_{2u} levels. The t_{2g} π bonding levels will be slightly lowered as a result of their rather stronger interaction with the metal t_{2g} levels, previously non-bonding. The net effect of the metal orbitals, and to bonding is thus to lower Δ, by raising the energy of the t_{2g} transfer an amount of charge from the occupied ligand π orbitals to the metal t_{2g} d orbitals as a result of the mixing.

WCl_6 is an octahedral compound of W(VI) similar to $W(CH_3)_6$, in which we predicted a transfer of charge to the methyl groups through the σ bonds. This will be accentuated in WCl_6, since Cl has a higher electron affinity than CH_3, but offset by the transfer of π electron density from the chlorine electrons to the metal t_{2g} d orbitals.

This prevents the tungsten atom carrying a ridiculously high charge and is in accord with Pauling's electroneutrality principle, which proposes it to be unlikely that any atom in a compound carries a charge of greater than ±1units. Support for this M.O diagram comes from the fact that the LFSF, Δ, is much less in complexes of ligands which are good π donors, such as chlorine, than those which are a donors (such as ammonia).

More graphically, the ion $[IrCl_6]^{2-}$ has five electrons in the weakly antibonding, mainly metal, t_{2g} shell. Clearly, one of these electrons must be unpaired, and if our arguments are correct, the small amount of chlorine π orbital mixed into this M.C. requires that this unpaired electron has a non-zero probability density on each chlorine atom.

The E.P.R. spectrum of this ion shows a considerable coupling of this electron with the chlorine nuclear spins, confirming that the electron is delocalised over the ligand atoms. The second case of metal-ligand π bonding is with ligands having empty π orbitals at a relatively low energy—a good example is the CO ligand which has low-lying unoccupied t_{1u} antibonding orbitals. These have the same symmetry as the first case, but now, as both the t_{1u} (ligand) and t_{1u} (metal p) orbitals are unoccupied, we need not worry about their interaction.

The metal t_{2g} orbitals will overlap with the ligand t_{2g} and, since the metal orbitals are lower in energy, the bonding orbital resulting will be mainly metal in character, but will involve a slight delocalisation of d electron density onto the ligands.

The metal t_{2g} orbitals are now bonding, and consequently Δ is increased by any π acceptance by the CO ligand. For the CO ligand, π acceptance corresponds to an occupation of the antibonding π* orbitals, and thus a weakening of the CO bond. This is shown by a fall in the CO stretching frequency from 2189 cm^{-1} to about 2000 cm^{-1} in metal carbonyl complexes. A good example of such a complex is tungsten hexacarbonyl $W(CO)_6$: the carbonyl group has a weakly bonding a orbital mainly localised on the carbon atom which overlaps well with the metal a orbitals. This orbital is doubly occupied, and consequently, there is a donation of charge from the 6 carbonyl orbitals to the metal, in the occupation of the 6 a M.O.s (e_g + a_{1g} + t_{1u}). There remain six electrons (from the tungsten) to be placed in the next M.O. the t_{2g} bonding.

This fills all the bonding M.O.s, giving a stable molecule, and also 'back donates' charge from the metal to the ligand—in total then, there is little transfer of charge between metal and ligands. This mode of bonding, in which charge is transferred in one direction by σ donation and in the opposite direction by π donation is known as synergic bonding. π bonding is thus possible in the presence of either (i) a π donor ligand, which mixes some metal t_{2g} character into its occupied π orbitals, decreasing Δ, and leading to a partial transfer of charge to the metal, or (ii) a π acceptor ligand, which mixes some of its empty antibonding π orbitals with the metal t_{2g} orbitals, increasing Δ, and resulting in a partial charge transfer to the ligands. An example of the first case is WCl_6; of the second case, $W(CO)_6$.

Having dealt with the electronic structure of these

complexes, we may now make some suggestions about their stability and reactivity. We mentioned that π accepting species increase Δ (as do strong, a donors), and consequently we would predict that complexes of such ligands would be low spin. Ligands which act as strong π donors (such as F^-) or which are weak a donors will tend to give high spin complexes: thus $[CoF_6]^{3-}$ is a high spin $3d^6$ complex, while $[Co(NH_3)_6]^{3-}$ is low spin, NH_3 being a stronger a donor, but a very weak π donor. Complexes with the t_{2g} orbitals occupied will be more stable if the ligands are π acceptors (the t_{2g} are then bonding) than with strong a donors such as CH_3 (t_{2g} non-bonding) or π donors (t_{2g} antibonding). This leads to the result that π accepting ligands stabilise lower oxidation states.

In complexes such as $[Co(NH_3)_6]^{2+}$ the weakest bound electrons are in weakly antibonding e_g* levels; their weak binding enables them to be removed with no real loss in stability of the complex, and $[Co(NH_3]^{2+}$ may be readily oxidised to $[C_6(NH_3)_6]^{3+}$ without breaking any metal-ligand bond.

The nonbinding or weakly bonding nature of the d electrons in many complexes accounts for the wide variety of oxidation states these species can show. In the case of manganese, it is even possible to remove electrons progressively from $[Mn(OH_2)_6]^{2+}$ (d^5) to $[MnO_4]^-$(d^0), the oxygen becoming a progressively stronger π donor as the protons are removed. The effects of the ligands in such systems are considerable: for example, to remove an electron from $[WF_6]^-$ (d^1) to give WF_6 (with structure similar to WCl_6) requires only 5.2ev.

The tungsten atom in such a complex definitely carries a slight positive charge, yet the energy to remove the lone electron from the t_{2g} orbitals is much less than the first ionisation energy of tungtsten (7.98 eV). Leands may also affect redox equilibria in solution, since they change the energies of the d orbitals. Co^{3+} and Mn^{3+} are rapidly reduced to the +2 state in the absence of stabilising ligands, but $[Mn(CN)_6]^{3-}$ and $[Co(NH_3)_6]^{3+}$ are stable and are formed by atmospheric oxidation from the +2 oxidation state complexes. The distinction between ionic and covalent bonding in transition metal complexes is not clear cut.

The closeness of tungsten 5d and carbon σ orbital energies make it unrealistic to take an ionic model (W^{6+}, $6CH_3$) as a starting point for discussion of $W(CH_3)_6$. The greater orbital binding energy of the donor lone pair of NH_3, and the lower oxidation state of the metal makes the ionic model more plausible for $(Co(NH_3)_6]^{2+}$.Oxidation to Co(III) will cause a drop in 3d orbital energy which will result in greater mixing with the nitrogen σ donor orbitals.

Tetrahedral coordination

Tetrahedral coordination is the second most common coordination of transition metal complexes. In pure tetrahedral geometry the M.O. diagram may be obtained as for the octahedral case: the ligand o orbitals transform as $a_1 + t_2$ (as for methane), and these can interact with the central atom s-orbitals (a_1) and d orbitals (t_2 set) and also (to a lesser extent because of their energy) the p orbitals (also t_2).

The e orbitals are not bonding and thus we expect a splitting between e and t_2 levels, with the metal t_2 d orbitals slightly antibonding and at a higher energy, The order of the two subsets of the d subshell is thus inverted in comparison with the octahedral case. Both e and t_2-levels are available for π bonding. Calculations, based on either a pure ionic model, or an angular overlap, model, and experiment show that the splitting between the e and t_2, d sub-shells is much smaller for tetrahedral than for octahedral complexes.

The ionic model predicts that if metal ligand bond lengths remain the same, Δ tet = 4/9 Δ oct and the experimental values found agree fairly well with this, although Δ tet is often slightly more than 4/9 Δ oct as the metal-ligand bond lengths are often shorter for tetrahedral complexes. The small value of Δ tet, and the fact that both sets of orbitals are equally affected by π bonding enable us to make certain predictions about tetrahedral complexes.

A is smaller than for octahedral complexes while the electronic repulsion parameters are the same—thus all tetrahedral complexes are high spin. There are no particular numbers of d electrons for which the tetrahedral arrangement is very strongly favoured, and in general an octahedralcoordination is more stable; however, with small central atoms, octahedral coordination may cause the ligands to overlap each other, and this generally produces an antibonding repulsive effect. The tetrahedral arrangement is common in such cases, e.g. $[CoC1_4]^=$, $[NiCl_4]^=$. Tetrahedral coordination is also found with d^{10} species such as $Ni(CO)_4$, $Ir(NO)(PPh_3)_3$ where the ligands may π-accept from both subsets of d orbitals.

For d^0 systems tetrahedral coordination is also found e.g. in $[WS_4]^=$ and OsO_4 where there is appre-ciable π donation from the ligands giving a high double bond character to the metal ligand bond. The d-d absorption spectra of tetrahedral complexes are often more intense than their octahedral counterparts. Referring to the M.O. diagram, we see that both metal d and p orbitals have t_2 symmetry and consequently the t_2^* M.O. may have some p character mixed in. The e→ t_2^* transition thus has some character, and is more intense since the d → p transition is allowed by the electric dipole mechanism.

Square planar complexes

There exist a considerable number of square planar

four-coordinate complexes of symmetry D_{4h}. The relevant orbitals transform in a peculiar way. If σ bonding only is considered then only the $d_{x^2-y^2}$ (b_{1g}) and d_{z^2} (a_{1g}) metal d orbitals will have suitable symmetry to overlap with the ligands.

We would expect the $d_{x^2-y^2}$ to overlap much more strongly as it is wholly directed at the ligands, while only the toroid 'doughnut' part of the d_{z^2} can overlap with the ligands. We thus expect the $d_{x^2-y^2}$ bIg orbital to be appreciably more antibonding. This explains the experimental observation that d^8 complexes which have the lowest four d orbitals occupied show a strong tendency to square planar geometry. The effect of π bonding are, as usual, smaller in magnitude. π acceptor ligands (e.g., Cl^-) energy of the σ non-bonding e_g and b_{2g} orbitals, and π donor ligands (e.g. Cl^-) raise their energy.

The effect on the d_{xy} (b_{1g}) orbitals is more pronounced as we would expect from the greater overlap with the ligands compared with the d_{xz}, and d_{yz} orbitals. The strongly antibonding nature of the b_{1g} ($d_{x^2-y^2}$) orbital is not affected by π bonding. There are two useful comparisons to be made with the square planar M.O. diagram: firstly, if we consider an octahedral complex in which the two ligands on the z axis are progressively removed and the xy plane ligands compressed, then, if the ligand is only a weak π donor, the $d_{x^2-y^2}$ will rise in energy, the d_{z^2} will fall (the or bonding effect), and the d_{xy} orbital will rise slightly as the xy plane ligands move in (the π effects) while the d_{xz} and d_{yz} will not be greatly changed.

Secondly, if we had considered squashing a tetrahedron to give a square of ligands, then the t_2 subset, which are antibonding and directed at the ligands for a tetrahedron will fall in energy, while the overlap with the $d_{x^2-y^2}$ will rise sharply, with an increase in energy. The

d_z2 will not be greatly affected. We have remarked that square planar geometry is favourable almost uniquely for d^8 species — we may now note that it will only be favourable if the ligand field splittings are sufficiently great to overcome the pairing of electrons necessary to leave the $d_x2—_y2$ b_{1g} empty.

Thus square planar complexes are typically strong field complexes. Thus $Ni^{2+}(3d^8)$ forms octahedral $[Ni(H_2O)_6]^{2+}$ and tetrahedral ($[NiCl_4]^{2-}$) complexes with weak field ligands, and with stronger, more covalent ligands gives planar complexes ($[Ni(CN)_4]^{2-}$). Copper (II) ($3d^9$), which would be expected to show all appreciable Jahn-Teller distortion in octahedral geometry, also forms some planar complexes. There are a small number of nickel complexes of the type NiX_2L_2, where X is a halogen and L a phosphine or airline ligand, which exist in tetrahedral and square planar forms, or which change structure with temperature. This seems to be due (in some cases) to a balance of electronic effects and in others to the crowding and steric hindrance of bulky L ligands in the square planar form.

Participation off orbitals in bonding

The lanthanides and actinides both have partially filled f subshells, and one might expect it to be necessary to derive molecular orbitals involving the f orbitals. In fact there is very little evidence for participation of the 4f orbitals in lanthanides in chemical bonding. Presumably the 4f orbitals are not sufficiently diffuse to give a good overlap with ligands. The lanthanides have an exclusively ionic chemistry and the 4f orbitals may be regarded as nonbinding.

The 5f orbitals of the actinides might be expected to be more diffuse, and consequently more likely to overlap with the ligands. The early actinides do show rather more covalent character but the bonding in their compounds is by no means well under stood, and is complicated by the small energy differences between 5f, 6d, 7s and 7p levels. The most conclusive evidence comes from the study of the organometallic complexes formed by uranium, neptunium, and plutonium with cyclo-octatetraene, C_8H_8.

Organometallic compounds

Organometallic compounds are almost invariably highly covalent, and the bonding electrons highly delocalised. Furthermore, the symmetry is generally much lower than in the complexes, in few cases is it possible to draw an M.O. diagram for the whole molecule, and discussion of electronic structure tends to concentrate on the character of a particular metal ligand bond. If it is impossible to draw M.O. diagrams for the molecule as a whole, it has been found possible to explain the stability of a large number of organometallic compounds by electron counting methods. It was noticed about 50 years ago that a large number of stable carbonyls had 18 electrons in the metal ligand bonding system.

We have already seen an example of this with $W(CO)_6$—there were 6 pairs of electrons in the (mainly ligand) a orbitals and 6 electrons in the (mainly metal) π orbitals. Another example would be $Fe(CO)_5$ where each CO is considered to donate 2 electrons to the metal ligand bonding, and the iron atom has 8 electrons in its valence shell ($3d_6 4s_2$), again giving a total of 18. Further support comes from the observation that the unstable $V(CO)_6$ (17 electrons) readily takes up an electron to give $[V(CO)_6]^-$ 18 electrons. The sum of the electrons on the metal and the electrons donated by the ligands should equal the atomic number of the inert gas next after the metal in the periodic table or the number of electrons in the valence shell of the metal plus the number contributed by the ligands should equal 18.

Such compounds have a greater stability than equivalent compounds with a different number of electrons. This 'effective atomic number' or '18 electron' rule is essentially a rule of thumb, and is unreliable for ionic complexes, but covers a very large percentage of organometallic compounds of transition metals, irrespective of their symmetry. One or two tentative explanations of the success of the rule have been proposed, of which the most reasonable is that the metal has nine atomic orbitals in its valence shell, (five d, one s and three p), and that the most stable bonding arrangement will be for it to form nine bonding and nine antibonding orbitals, with all the bonding orbitals filled—corresponding to 18 electrons in the valence shell. Some force is given to this argument by the fact that, for the first row elements, the rule works best in the middle of the 3d series where 3d, 4s and 4p orbitals are closest in energy.

Table 2

The number of electrons donated by variuos ligands

	Number of e^-	Examples
Covalent bond	1	CH_3, H, Cl^-
Lone pair	2	terminal Co, NH_3,
Oliefin	2	C_2H_4, C_2F_4
Allyl group	3	C_3H_5
Nitrosol	3	NO
Diene	4	C_4H_6
Aromatic ring	6	C_6H_6, $C_5H_5^-$

The majority of the exceptions to the 18 electron rule are 16 electron compounds, many of them square planar d^8 complexes where, as has been mentioned, one of the d orbitals (the d2) is non-bonding. If the 18 electron rule is accepted (and it is by no means the only electron counting rule to work for no especial reason) it is clearly necessary to decide how many electrons each ligand can give. The numbers for a variety of common ligands are given in Table 2.

Carbonyls

There are a very large number of carbonyls, and substituted carbonyls and we note here only a few of interest. $Mn(CO)_5$ is a π electron system and might be expected to be unstable; in fact it is unknown, preferring to form a Mn-Mn single bond to give a dimer $Mn_2(CO)_{10}$, each Mn atom donating one electron to the valence shell of each other (i.e. $Mn(CO)_5$ acts as a an electron ligand) to give two electron systems. [$Mn(CO)_5$ can however, be prepared, as may [$Mn(CO)_6$. Alternatively, a series of one electron donors may occupy the sixth coordination site of the $Mn(CO)_5$ species to give an 18 electron system e.g. $CH_3Mn(CO)_5$; $HMn(CO)_5$; $ClMn(CO)_5$. $CO(CO)_4$ is again a $17e^-$ system and is unknown—however the dimer $CO_2(CO)_8$ exists. Rather than form a simple $(OC)_4CO\text{-}CO(CO)_4$ structure it adopts a structure with two bridging CO groups and a metal-metal bond. The electron count is now (for each cobalt atom):

Cobalt	9
3 terminal CO	6
2 Bridging CO	2
1 From the other	
Co (metal-metal bond)	1/18

The X-ray crystal structure shows that the CO bond length is slightly greater for the bridging carbonyl groups than for the terminal groups. This implies that the CO bond is slightly weaker in this case, and this is confirmed by the infrared stretching frequency. Terminal carbonyl groups generally absorb near 2050 cm^{-1} (free CO is 2143 cm^{-1}) while bridging groups absorb typically at

1850 cm^{-1}—this is in fact not too far from an organic ketone absorption showing the strong covalent nature of the M-CO-M linkage. Triply bridging carbonyls are known (e.g. $Rh_6(CO)_{16}$), and absorb at lower energy (1750 cm^{-1}). The weakening of the CO bond results from the withdrawal of electrons from cr bonding or non-bonding orbitals and the donation of electrons into their antibonding orbitals.

Thus the metal-ligand bond is strengthened at the expense of the internal bonding of the ligand. This is extremely common in organometallic chemistry. It may be noted that when the CO groups in a carbonyl are substituted by a weaker or non acceptor (e.g. pyridine (C_5H_5N), ethylendiamine) the infra-red frequency falls further, indicating that the CO now π accepts a greater charge. CO frequencies have been used extensively as an indication of the bonding nature of other ligands in the complex.

Analogues of carbonyls

There are several ligands which have a electronic structure similar to carbon monoxide, notably NO^+, N_2, $C{\equiv}CH^-$, CN^- all iso-electronic with CO, iso-cyanides (RN=C) and the CS molecule, unknown except as a ligand. All are assumed to bond in a roughly similar way, acting as or donors and π acceptors. Nitrosyl complexes form the greatest number of analogues to carbonyls, and incidentally also give us further information on the nature of the bonding of these groups. If we take a simple example, $(Ph_3P)_2Ir(NO)(CO)$ the NO stretching vibrations absorb at 1645 and 1660 cm^{-1}.

Free NO is paramagnetic with one electron in the π* orbitals and absorbs at 1860 cm^{-1}. The lower frequency in the complex suggests that there is further π acceptance into the π*. Nitrosonium salts (e.g. $NO^+ BF_4^-$) exist and absorb at about 2200 cm^{-1}. The iridium complex is diamagnetic showing that, the unpaired electron is sufficiently involved in bonding to pair with the unpaired electron on the iridium. A feature of NO complexes is the variation in the metal N-O bond angle. In most nitrosyls this is close to 1800, but a small number have been found where the bond angle is close to 1200. In such cases (e.g. $[Ir(CO)CI(PPh_3)_2(NO)]^+$) the NO stretching frequency is generally rather lower than for complexes with linear metal-NO groups.

The metal-nitrogen bond in such cases is rather longer than usual, and it has been suggested that the N atom is sp_2 hybridised (giving the bond angle of 1200) and rather closer to NO^-. This makes the complexes obey a 16 electron rule rather than an eighteen electron rule, but as they generally involve a d^8 metal (Ir(I), Ru(O) which gives many 16 electron complexes this is perhaps not surprising. Molecular nitrogen forms a fairly large number of complexes in which the linear mode of bonding is similar to CO.

The stretching frequency is reduced from 23-31 cm-1 to between 1900 and 2150 cm^{-1} and bond length increase for example the bond length in $Co(PPh_3)_3(N_2)H$ is 1.112 A compared with 1.098 A in free N_2. It appears to be a rather weak a donor, but a strong π acceptor, and molecular nitrogen in such complexes can sometimes be reduced to ammonia or amines under relatively mild conditions. The bonding of cyanide ion and isocyanides is very similar to that of carbon monoxide as the similarity of their electronic structure would suggest.

Several other simple molecules, in particular phosphines and arsines (such as $P(C_6H_5)_3$, $AsEt_3$) may be substituted for CO in metal carbonyls. Evidence suggests that they also act as u donors and π acceptors, but the π electrons are generally held to be accepted into the empty d orbitals of the ligating atom.

Olefin complexes

The olefin complex $K^+[PtCl_3(C_2H_4)^-$ prepared by Zeise has been known for nearly 150 years, but a satisfactory bonding description of the complex was only given in the early 1950's. The structure of the ion is shown in Fig.5. The platinum shows square planar coordination typical of a third row transition metal d_8 complex.

The sketch of the π orbitals of the ethylene molecule shows that the π bonding orbital ϕ_1 has π symmetry about the Pt-olefin (z) axis, and the π antibonding orbital ϕ_2 has σ symmetry with respect to this axis. Thus we have again a ligand with a full a and empty π orbitals. The model for the bonding proposed by Dewar and Chatt is that there is a transfer of charge from the ϕ_1 orbital to the platinum accompanied by back donation from the platinum d_{xz} orbital into the orbital ϕ_2 —in other words the familiar synergic bonding found in carbonyls.

The Dewar-Chatt model is generally accepted as the most satisfactory explanation of the bonding of olefins. For the purposes of electron counting, the ethylene ligand is regarded as a 2 electron donor. The effective transfer of charge from the π bonding orbital ϕ_1 to the π antibonding orbital ϕ_2 would be expected to weaken the C=C double bond—this may be shown by the drop in the C—C stretching frequency (1551 cm^{-1} in Zeise's salt, 1623 in ethylene) and a lengthening of the C-C bond (by up to 0.1 Å).

Electron withdrawing substituents (such as fluorine or cyanide) on the ethylene would be expected to increase the π acceptance by ϕ_2, and weaken still further the C=C bond; in fact $C_2(CN)_4$. Complexes show an 'olefinic' carbon-carbon bond length little shorter than a carbon-carbon single bond. Alkenes coordinated to transition metals are much more susceptible to hydrogenation; this is the basis of many catalysts.

From a molecular orbital point of view, the transfer of charge from ϕ_1 and to ϕ_2 is Similar to a photochemical activation involving the jump of an electron from π to π^*. The excited states of ethylene and similar molecules are frequently found to be bent, and it is found that the C_2X_4 species is no longer planar, and that the two CX_2 fragments are bent away from.the metal atom. This has led to the suggestion that coordinated olefins are similar to cyclopropanes, and are essentially σ bonded. One of the more remarkable features of coordinated ethylene is the rotation of the C_2H_4 molecule about the metal—C_2H_4 bond axis.

The nuclear magnetic resonance spectrum of the compound $(C_5H_5)Rh(C_2H_4)_2$ shows that the ethylene rotates about the Rh-alkene bond axis with a rotational barrier of about 60 kJ/mole. Presumably the destruction of π bonding with the d_z orbital on rotation is compensated by π bonding with the d_{yz} orbital.

Electron deficiency

As synthetic chemistry progressed, and as studies of the structures of compounds became more common, it became clear that there were several compounds for which such simple bonding schemes could not be proposed, as there were not enough electron pairs to give the correct number of bonds. Such compounds were dubbed electron deficient—a little unfairly as the deficiency ties with the bonding theory, and the name tends to imply that such compounds would be good electron acceptors (or oxidising agents), which is by no means always the case. The accepted definition of an electron deficient compound is one in which m atoms are held together by less than $2(m\text{-}1)$ electrons, or one for which no two centre electron pair bonding scheme can be drawn. From the point of view of the more general molecular orbital theory we have been using, there is nothing extraordinary about electron deficient compounds, but they exhibit a sufficient number of rather unusual properties to justify a separate discussion. The classic example of electron deficiency is diborane (B_2H_6) Chemically, the terminal BH bonds show no peculiarities, and we may simplify our discussion without any real loss by assuming them to be localised electron pair bonds.

Furthermore, the terminal H—B—H angle of 122° suggests that we may regard the terminal H as bonding with sp^2 hybrids. For the bonding involving the other bridging hydrogens we have thus to consider the sp^2 hybrids not involved in terminal BH bonding, p orbitals not involved in the sp^2 hybrids (ϕA_2 and ϕB_2), and the two hydrogen orbitals (ϕH_1 and ϕH_2). We thus have 6 orbitals; these may be split up by their symmetry properties (the point group is D_{2h}) but we can take a short cut, and notice that ϕA_2 and ϕB_2 and the combination ($\phi H_1 + \phi H_2$) are symmetric with respect to rotation about the B-B axis, while ϕA_2, ϕB_2 and ($\phi H_1 + \phi H_2$) are antisymmetric.

These two sets will thus be mutually orthogonal. Within the sets, the three orbitals will give bonding, non-bonding and antibonding molecular orbitals. Boron has three electrons in its valence shell, and hydrogen has only one; the total number of valence electrons in the molecule is thus 12. Eight electrons are taken by the simple B—H terminal bonds, leaving four for the bridging groups.

These four electrons will fill the two bonding orbitals of the bridging M.O. described above, and the description of the molecule involves only bonding orbitals being filled—consequently the molecule is stable. The bridging hydrogen system of the molecule is delocalised over 4 atoms (2 hydrogen and 2 borons). A suitable combination of the two occupied bonding orbitals gives 'localised' bonds having maxima of electron density along the two B—H—B curves. These localised bond are three centre, two electron bonds. The importance of the three centre two electron bond is that it enables a stable structure with one electron pair forming two 'bonds' to be proposed; this is impossible according to classical electron pair bonding theory. If a full M.O. treatment of the molecule is undertaken one arrives at exactly the same picture of the bonding, although one of the unoccupied 'non-bonding' M.O. of the bridge system is appreciably antibonding.

There is nothing fundamentally startling in the M.O. description of the bridge bonding system. However, one electron pair delocalised over three centres and providing two formal bonds produces a weaker bond than two 2 centre 2 electron bonds. Wade gives the estimated energy of the terminal B—H bonds as 3 75 kJ/mole and of a B—H—B bond as 450 kJ/mole —much less than twice the 2 single B-H bond value.

The B—H—B bridging system is easily cleaved by a Lewis base which has a pair of electrons which can overlap well with an acceptor molecule and which is not too strongly bound. Thus the reaction

$$B_2H_6 + 2Me_3N \rightarrow 2Me_3NBH_3$$

is very favourable, involving the interaction of 2 lone pairs and two 3 centre 2 electron bonds to give four 2 centre 2 electron bonds. All boron hydrides act as Lewis acids, and are decomposed by bases. The instability of these compounds is further shown by their positive heats of formation showing them to be unstable with respect to decomposition into boron and hydrogen.

The results of the extremely difficult preparative and structural studies of the boron hydrides undertaken by many chemists (notably Stock and Lipscomb) showed that many more complex boron hydrides existed, frequently sharing the same type of B—H—B bridged bonds, and forming cages and rings with boron-boron bonded skeletons. It was also found that many of these compounds showed boron-boron bonds which could not be explained according to classical electron pair bond theory.

Two types of three centre two electron bonds for boron polyhedra were proposed: one in which three 'a type' boron orbitals overlapped to give bonding, non-bonding, and antibonding molecular orbitals, and the second in which two boron cr type and one boron p orbital overlapped to give 3 M.O. Using these three types of three centre bond it was possible to draw a bonding scheme for many boron hydrides. Hexa-borane(10) B_6H_{10}—its structure consists essentially of a pentagon of boron atoms with a BH group above the centre of the pentagonal plane.

This approach has the disadvantage that sometimes several bonding schemes can be drawn for the same compound and that the molecule often has a higher symmetry than the bonding pattern would predict, as for example in the anions $B_6H_6^{2-}$ (octahedral) and $B_{12}H_{12}^{2-}$ (icosahedral). This problem can be resolved by a full M.O. treatment allowing for delocalisation throughout the molecule.

Several examples of this are given in the excellent review by Longuet Higgins and we shall only give a simple example here. $B_6H_6^{2-}$ has an octahedral structure with a B—H unit at each apex of the octahedron. In total there are 18+6+2 = 26 valence electrons of which 12 will be involved in simple BH bonds, leaving 14 to hold the octahedron together. The boron hydrogen bond is assumed to involve a sp hybrid, leaving one sp hybrid directed towards the centre of the octahedron and two p orbitals in the plane perpendicular to the sp hybrid axis.

The sp hybrids directed to the centre of the octahedron will form one strongly bonding M.O. (of symmetry a_{1g}) of the form $(\phi_1 + \phi_2 + \phi_3 + \phi_4 + \phi_5 + \phi_6)$ where ϕ_i is the sp hybrid of the i-th boron atom. The other M.O. formed by these orbitals will be antibonding. There are 12 p orbitals perpendicular to these sp hybrids which will produce 6 antibonding and 6 bonding M.O.

The total of bonding M.O.s is thus 7—corresponding to 14 electrons, the observed number of skeletal bonding electrons (i.e. the number required to hold the boron cage together). A similar treatment for B_4Cl_4 (a tetrahedron of B—Cl units) yields an equally satisfactory picture of the M.O. diagram if it is assumed that the Cl atom can act as a donor of π electron density. These full M. O. treatments are satisfactory for species with high symmetry, but much more difficult for those species with lower symmetry. As more preparative and structural work appeared, certain general features of the boron skeleton became apparent.

The boron skeleton always formed either the most spherical triangular faced polyhedron, or a fragment of such a polyhedron (for example: a square based pyramid is an octahedron with one vertex missing; a square planar structure is an octahedron with two transvertices missing). The form of this polyhedron was governed by the number of skeletal bonding electrons. A species with n boron atoms in the cluster would form a closed polyhedron with n vertices if it had $2n + 2$ skeletal bonding electrons: if it had $2n + 4$ skeletal electrons it formed a polyhedron with $(n + 1)$ vertices with one vertex missing; if it had $2n + 6$ electrons it formed a polyhedron with $(n + 2)$ vertices, two of which were missing.

Chapter 3

Properties of Elements

Dalton published his Atomic Theory with its concept of the regrouping of small whole numbers of atoms of elements reacting. The atoms, as their name suggests, were indivisible and of masses characteristic of the elements involved. By Atomic Theory Berzelius determined equivalent masses and, having deduced the composition by atoms of many common compounds. Dobereiner was the first to group similar elements together and link this with their atomic masses.

Here are some of the analogies which lie noted: calcium, strontium and barium form a sequence of elements with similar properties and strontium (= 88) has an atomic mass which is the arithmetical mean of those of calcium (40) and barium (137). He stressed that similar arithmetical relationships, such as between the atomic masses of carbon (12), nitrogen (14) and oxygen (16), were meaningless without chemical similarities.

However, some elements which are similar do not form these sequences, or 'triads', and Dobereiner suspended judgment here Until further data were available. Such a group is manganese (55), iron (56), cobalt (59), nickel (59) copper (64) and zinc (65), which are elements with nearly the same atomic masses.

Cannizzaro, de Chancourtois

The first real attempt to correlate chemical properties and

atomic mass was made by de Chancourtois, who plotted atomic masses in a 45° spiral or helix on the surface of an upright cylinder. Similar elements were in the same vertical straight lines and were separated by 16 units of atomic mass. Listing elements in increasing order of the (at that time) new relative atomic masses, he noted that similar elements were separated by intervals of eight elements.

He called this observation the 'Law of Octaves': it was received with ridicule. Comparison, remembering that the inert gases had not been discovered at the time, shows that his table bears a striking similarity to the Periodic Classification although the order of the elements was disorganized by some incorrect values of atomic masses and by undiscovered elements. Part of the scheme advanced by Newlands is shown in Table 1.

Mendeleeff and Lothar Meyer

Attending the Congress of Karlsruhe influenced by the ideas of Cannizzaro were two men who would shortly write textbooks requiring the collation of the physical and chemical properties of elements and a detailed examination of atomic mass values: both were struck by the periodicity of properties with atomic mass. Neither seemed aware of the work of others in this field.

Table 1
Classification of the elements Newlands

	H	F	Cl	Co, Ni	Br
Order	Li	Na	K	Cu	Rb
of	Be	Mg	Ca	Zn	Sr
increasing	B	Al	Cr	Y	etc
atomic	C	Si	Ti	In	
mass	N	P	Mn	As	
	O	S	Fe	Sc	

Lothar Meyer was primarily interested in the

relationships obtained by plotting the magnitude of physical properties against atomic masses. We are chiefly indebted to him for the atomic volume curve of which a modern form, substantially the same as the original. The periodic value of the property is clearly displayed.

The *atomic volume* of a solid element is the volume of the molar mass.

$$\text{Atomic volume} = \frac{\text{molar mass}}{\text{density}}$$

The unit for atomic volume is $cm^3\ mol^{-1}$. Many other physical properties such as melting-point, boiling-point, coefficient of thermal expansion, and malleability, to name but a few, show similar types of curve. From these studies Meyer drew up a periodic table. Mendeleeff announced his system of classification. Basing his reasoning on chemical characteristics he listed elements in order of atomic masses and showed the periodicity of properties. By making predictions about the existence of hitherto undiscovered elements, which were confirmed within a few years, he quickly got his ideas accepted. His observations are summarized in the next section.

Mendeleeff's Periodic Law

The properties of the elements are a periodic function of their atomic masses. In reaching his general conclusion which we have called the *Periodic Law,* Mendeleeff noted that similar elements had either almost the same atomic masses, e.g. platinum (195), iridium (192), osmium (190), or values which showed a uniform increase, e.g. potassium (39), rubidium (85), caesium (133), the modern values having been given here, to the nearest integer.

Mendeleeff stressed that atomic mass determines the character of an element and the arrangement of groups brings out the differences in chemical nature and valency along each 12 series. The elements with light atoms were found to be generally widely distributed and to have clearly defined properties. He called these *the typical elements.* They are located in the three short periods. Hydrogen was recognized as unique, or 'typical of itself'. The group similarities are so striking that it tends to be forgotten that Mendeleeff also stressed the similarity of some adjacent elements.

Group VIII contains three sequences of triads to which Mendeleeff gave the name transitional or *transition elements.* This term has been extended to cover the sequence scandium (21) to nickel (28) in period 4 and the corresponding elements of the following periods. Lanthanides and actinides are discussed later. In the original Mendeleeff Periodic Table, Group O was missing because none of these elements was known. When the work of Rayleigh and Ramsay resulted in the discovery of a gas, which was chemically inert and quite unlike any known element, and which was therefore called argon (Greek: *argos* = inert or idle), other members of the family were immediately sought and found.

For some elements the positions allotted in the table and those apparently indicated by the study of chemical properties were not in agreement using the values of the atomic masses assigned to the elements at that time. Mendeleeff considered that the wrong multiple of the equivalent had been used in fixing the atomic masses of beryllium, indium and uranium. He adjusted the values so that the elements fitted into his scheme and subsequent experiments duly confirmed his predictions.

Minor adjustments of atomic masses have also been made with success but a major weakness of the Mendeleeff classification is made manifest in this connection. In the modern form of the Mendeleeff Table it will be seen that four pairs of elements, positioned by their chemical characteristic, are in the reverse order of atomic mass. These elements with atomic numbers and atomic masses are listed in Table 2. The root of the problem lies in the fact that the atomic mass of an element is not a fundamental quantity by which to classify the element.

The difficulty may only be resolved by the study of atomic structure where a table of rather different shape will be preferred. In addition to the lack of emphasis on the unique nature of hydrogen, some elements in the short form of the Periodic Table appear in positions which are at variance with their chemical properties. Because of the double series in the long period, transition metals in the wider sense, such as copper, silver and gold (in this book called 'associated with transition elements'), are placed alongside the metals of Group 1, potassium, rubidium and caesium, to which they bear no resemblance.

The original transitional elements of Mendeleeff were placed in triads in Group VIII. The term 'transition element' will be defined in terms of atomic structure later. The rare earth elements, all very similar chemically, and now usually called the lanthanides, proved even more difficult to place on the basis of atomic masses and are seen to be confined to one place in the Mendeleeff short Table.

We will now review the main discoveries and theories which have resulted in a knowledge of atomic structure and finally the classification of elements by the electronic configuration of their atoms.

Divisible atom

According to Dalton's Atomic Theory, the ultimate particle of matter was the atom, indivisible and characterized by its mass, all atoms of a given element being identical and different from atoms of other elements. It had been suggested by Prout, that the hydrogen atom was the fundamental unit from which all other atoms were made.

Table 2

Atomic mass anomalies in Mendeleeff's classification

Element	*Symbol*	*Atomic number*	*Atomic mass*
Argon	Ar	18	39.948
Potassium	K	19	39.102
Cobalt	Co	27	58.9332
Nickel	Ni	28	58.71
Tellurium	Te	52	127.60
Iodine	I	53	126.9044
Thorium	Th	90	232.038
Protactinium	Pa	91	231.036

While current values of atomic masses seemed to support: this hypothesis, the more accurate determinations. Experiments now show that the concept of indivisibility and uniqueness of mass of atoms must be modified. Yet the effectiveness of the Classical Atomic Theory in stressing the fundamental nature of the atom in chemical reactions has not been diminished with the greater understanding of atomic phenomena.

Atoms of certain elements undergo spontaneous disintegration with the emission of particles and rays. This is known as radioactivity and was discovered by Becquerel. When cathode rays, which are high-speed electrons, collide with a solid object, X-rays are produced. These have characteristic properties notably, they produce fogging of photographic plates wrapped in black paper to exclude ordinary light, excite fluorescence in certain compounds, render gases electrically conducting by causing ionization of their molecules, and pass through thin sheets of substances opaque to ordinary light. Investigating the connection between X-rays and fluorescence, Becquerel discovered that uranium and its compounds, without previous treatment of any kind, produce the same effects as X-rays. He concluded that the element uranium spontaneously emits radiation. The intensity of the radioactivity, which could be measured by the rate of discharge of a charged electroscope due to ionization of the contained air, depended only on the quantity of element, uranium, present.

The radioactive decay of an element proceeds spontaneously and is independent of changes in the physical conditions. Temperature change has no effect on radioactive decay which is in marked contrast to the influence seen on the rates of ordinary chemical reactions. Radioactivity of certain uranium minerals, notably pitchblende (U_3O_8), exceeded that calculated for its uranium content Marie Curie with her husband, P. Curie, looked for other radioactive elements in these minerals. This led to the discovery of polonium and radium.

During the next year, Debierne, another colleague of Marie Curie, discovered actinium also in uranium minerals. In addition, many other elements have been discovered which exhibit radioactivity to some degree besides those cases of artificial radioactivity produced in laboratories and atomic piles. The Becquerel rays are of three different kinds, two named by Rutherford in α (alpha) rays and β (beta) rays, and the third γ (gamma) rays, following the work of P. Curie and Villard.

The three types of ray may be separated by using a powerful magnetic field, when β-rays are deflected in accordance with their negative charge, α-rays in the opposite direction and to a much less extent, and γ-rays pass undeflected. Ramsay and Soddy, showed that helium is Continuously produced during the decay of radium.

Rutherford and his co-workers proved that α-rays are positively charged helium atoms. He^{2+}, of mass 4 (on the relative atomic mass scale) and each bears a double positive charge, i.e. twice the magnitude of the charge of the electron but positive. β-rays have been shown to be fast moving electrons and are also called β^--particles. γ-rays are similar to X-rays, but with wave- lengths in the range approximately 10^{-12}M to 10^{-10}m, just shorter than those for X-rays, endowing them with even greater penetrating power. The Theory of Radioactive disintegration was advanced by Rutherford and Soddy.

The spontaneous disintegration of atoms with the emission of radiation results in the formation of atoms of new elements, which may also be radioactive and break down further. The rate of decay of radioactivity of each element will be characteristic of that element. The rate of decay follows an exponential law which means that at any instant the rate of decay is proportional to the intensity of the activity at that instant.

The intensity of the radioactivity will be reduced with time but will not disappear. The rate of decay, being independent of physical and chemical conditions, and being a characteristic property of the element undergoing radioactive changes, is usually expressed in terms of the 'halflife period' which is the time that elapses for the radioactivity to decay to one half of its value from a given instant.

Radioactivity, $A = \frac{-dN}{dt} = \lambda N$

(where N = number, t = time, λ = decay constant)

$$\text{Re-arranging, } \frac{+dN}{dt} = -\lambda \, dt$$

$$\text{and integrating, In } (N_t/N_0) = -\lambda t$$

where N_0 = number at $t = 0$ and N_t at time t)

$$\text{or } N_t = N_0 e^{-\lambda t}$$

and if $t_{1/2}$ represents the half-life period of the process

$$\text{In } 1/2 = -\lambda t_{1/2}$$

$$-2 = -\lambda t_{1/2}$$

$$t_{1/2} = \frac{In2}{\lambda} = \frac{0.693}{\lambda}$$

In calculations, N_t represents the number of atoms which have not disintegrated at time, t. The unit for measuring.

$$A = \left(= \frac{dN}{dt}\right)$$

is s^{-1} (= Hz).

Atomic nucleus

When a stream or pencil of a-particles, travelling at high speed from a suitable source, such as radium, is directed towards a thin piece of metal foil, a limited number of particles are deflected through wide angles while the majority suffer no appreciable deflection at all. In the first experiment, copper, silver, gold and plantinum of thickness about 0.0005 mm (5×10^{-7} m) were used and the α-particles were detected by the scintillations of light which each produced on collision with a zinc sulphide screen.

As an illustration, using gold foil 0.0004 mm (4×10^{-7} m) thick, the average deflection of the α-particles was less than 7°, but one α-particle in 20 000 was deflected through 90° or more. The experiments on the scattering of α-particles were directed by Rutherford. He concluded that the reflection of each a-particle was due to a single collision with an atom of metal. Because atoms contain (negative) electrons there must also be a positive charge to give electrical neutrality. That relatively few α-particles are deflected widely indicates that this charge and the greater part of the mass is concentrated in a small nucleus.

Electrons are much too light to cause deflections and most of the α-particles travel straight on, so it is assumed that the greater part of an atom consists of empty space. A picture emerges of a central, dense, small, positively charged nucleus surrounded at a relatively great distance by a sort of planetary system of electrons. Rutherford supposed that their high speed prevented electrons from spiralling into the nucleus but this is not so.

A more accurate treatment requires the use of wave, or quantum, mechanics. The charge on the nucleus is about one half of the atomic mass. Moseley bombarded various elements, or compounds where it proved difficult to use the elements, with cathode rays in a discharge tube and showed that the resulting X-rays had characteristic frequencies.

The square root of the frequency of a characteristic line in each spectrum was directly proportional to the atomic number of the element. Van de Brock suggested the hypothesis that the atomic number equalled the nuclear charge as determined by scattering experiments. From his experiments, Moseley concluded that the atomic number was a fundamental quantity for an element and probably equalled the positive charge on the nucleus, the unit being the charge of an electron.

There was no such precise relationship as indicated above when atomic mass was substituted for atomic number. This shows the decline in importance of atomic mass in element classification and the exceptions in the order based on atomic mass are shown in a new tight.

Chadwick determined the nuclear charge of copper, silver and platinum directly using a refinement of the scattering technique and the results were in excellent agreement with the atomic numbers. Now the α-particle is a positively charged helium atom, and helium has a higher atomic mass than hydrogen.

Therefore, a positively charged particle which is smaller than the α-particle was expected. The hydrogen atom was though to consist of a nucleus of the hydrogen atom was called the proton by Rutherford. He had produced these particles during the bombardment of nitrogen by α-particles. Protons could also be produced by the similar bombardment of other atoms and were assumed to be a constituent of all atoms.

Protons were also detected in the positive rays of the discharge tube. Hence, the atomic number of an element is equal to its proton number, the terms being interchangeable. Rutherford postulated the existence of an uncharged particle, which he called the neutron, of the same mass as the proton and forming part of the nucleus.

Structure of the atom

An atom consists of a nucleus, which is relatively small and dense, surrounded at relatively great distances by electrons, which are of negligible mass. The simplest atom, that of hydrogen, consists of one proton, which has a diameter of approximately 10^{-15} m and one electron moving at a distance around it.

The diameter of the atom is approximately 10^{-10} in. More complex atoms have diameters of the same order with the diameter of the nuclei being between 10^{-15} in and 10^{-14} in. As the diameter of an atom is approximately 10^{5} times the diameter of the nucleus, the atom is seen to contain mostly empty space. If the nucleus were magnified to 20 mm diameter, the electron would be at a distance of about one kilometre from it! The detailed pattern taken up by the electrons, upon which the properties of an element depend, will be described shortly. A simple picture rather like a miniature solar system in which the sun represents the nucleus and the planets are the electrons moving in orbits, is sometimes very helpful but not accurate.

The atom should be visualized as a small, dense, positively charged core surrounded by a diffuse cloud of moving electrons, the particle nature of which cannot be distinguished in the picture while the density of negative charge represented by this cloud falls off with increasing distance from the nucleus. It is difficult to define an exact boundary for the atom. The nucleus consists of protons and neutrons sometimes collectively known as *nucleons,* which account for nearly the whole of the mass of the atom.

The nucleus has a positive charge due to the protons and in the neutral atom the number of external electrons is equal to the number of protons, i.e. the proton number or atomic number. Different elements may be successively constructed from the simplest atom, hydrogen, which consists of one proton to the nucleus and one external electron at each step. Neutrons provide the ballast and make up the mass of the atom. For convenience, the atomic mass is sometimes rounded off to the nearest integer and called the Mass Number or Nucleon Number.

The Atomic Number (Z) of an element is equal to the following:

1. The value of the positive charge on the nucleus of an atom of the element, the unit of charge being the charge of one electron.
2 The order of the element in the Periodic Classification.

3 The number of electrons in a neutral atom of the element.

Z is also called the Proton Number.

The Mass Number (A) of an element is the relative atomic mass rounded off to the nearest whole number. A also represents the Nucleon Number.

An Element is a substance having the same number of protons in the nucleus of each atom of the element, the number being characteristic of the element.

By subtracting the proton number (Z) from the nucleon number (A), the number of neutrons (N) in the respective nucleus is obtained,

$$N = A - Z$$

We have seen that the atomic mass of an element is much less important than was formerly believed, but why are the values sometimes so far from being approximately integral as would be expected from structures composed of neutrons and protons, each of mass approximately unity? Why is the atomic mass of element No. 17 (chlorine) 35.453? Soddy suggested that the atoms of a given element could differ in mass. This came, from the study of radioactive decay where successive elements are formed from uranium and thorium. The non-radioactive end-products are chemically identical with lead but of atomic masses 206 and 208 respectively while the atomic mass of lead is 207.1.

The different forms of an element were called *isotopes* (Greek: *isos* and *topos* = equal place, i.e. in the Periodic Table). As a result of the preliminary experiments by J. J. Thomson and the later work of great precision by Aston, who had developed the mass' spectrograph by 1919, it was clear that elements not associated with radioactive decay could also consist of atoms of different masses. Neon was shown to consist mainly of atoms of mass number 20 with some of mass number 22, and chlorine of mass numbers 35 and 37. The chemical atomic mass is seen to be an average for the various atoms making up the sample used. The number of naturally occurring isotopes of successive elements fluctuates from as many as ten for tin down to just one for iodine, sodium, arsenic and others.

Isotopes are forms of the same element which differ in relative atomic mass. At one time it was thought that except for elements resulting from radioactive decay, the isotopic ratio for an element would be constant because of the thorough mixing of samples during the cosmic creation of this planet.

Precise measurements show that while this is correct for some elements, there is a smaller isotopic variation for others. Chlorine and silicon show no variations but this is not the case for sulphur, boron, carbon, silicon, hydrogen, krypton, xenon or nitrogen, and a slight variation in the isotopic composition of atmospheric oxygen and oxygen combined in water has been demonstrated.

Stewart pointed out that different elements may have isotopes of the same mass and he called these *isobares* or *isobars* (Greek: same weight). Thus, there are naturally occurring isotopes of argon, potassium and calcium all with mass number 40. We seem to be very near the early hypothesis advanced by Prout because now that we believe the different isotopes of an element are composed of neutrons and protons, the mass of an atom should equal the total mass of the constituent particles.

However, for a given isotope there is a small but

significant difference in mass. This is ascribed to the *packing effect.* When isolated nucleons are assembled to form the atomic nucleus of an atom, the energy of formation comes from the mass change recorded.

Einstein explained this equivalence of mass (m) and energy (E) in terms of the equation, E = mc^2 where c = velocity of light (3×10^8 ms^{-1}) so that a minute loss of mass results in an enormous release of energy because the c^2 term introduces a factor of 9×10^{16}. The effect is small and (A - Z) still gives the number of neutrons. This may be illustrated. Helium is virtually 100% $^{4}_{2}$He and may be built up with two hydrogen atoms, each containing I proton and I electron, and 2 neutrons:

Relative atomic mass calculated

$$= (2 \times 1.008) + (2 \times 1.009) = 4.034$$

Relative atomic mass by experiment = 4.004

Sodium is 100% $^{23}_{11}$Na and may be built up from 11 H atoms, each I proton and I electron, and 12 neutrons.

Calculated relative atomic mass

$$(11 \times 1.008) + (12 \times 1.009) = 23.196$$

Relative atomic mass by experiment = 22.990

In the symbols $^{4}_{2}$He and $^{23}_{11}$Na, the upper figure refers to the nucleon (mass) number (A) and the lower figure to the proton (atomic) number (Z) defined earlier.

The determination of atomic masses by the mass speedometer

Isotopic masses and the abundance ratio of the isotopes of elements may be determined, and hence the *weighted mean* unitless relative atomic mass calculated (or molar mass/g mol^{-1}), taking into account the proportions of the atoms of different mass present. A mass *spectograph* generally used for individual mass determinations, delivers a photographic record while a mass *spectrometer* incorporates a meter to measure ion current and is used to determine isotopic abundance ratios.

Today mass spectrometry is concerned with the determination of relative molecular mass (molar mass/g mol^{-1}) and the deduction of molecular structures, particularly of organic compounds. The principle of the mass spectrometer is described here. Positive ions are commonly produced by electron bombardment,

$$M + {}^{0}_{-1}e^{-} \rightarrow M^{+} + 2\ {}^{0}_{-1}e^{-}$$

but ultraviolet irradiation, sparking or field emission from a surface upon which the sample has been absorbed may be used. The element, or a suitable compound, is introduced as a gas under very low pressurc (1×10^{-4} Pa ($N\ m^{-2}$)) to avoid interference from other entities as far as possible. Molecules may ionize to ions with one or more charges, or simply disintegrate.

The positive ions are accelerated by a powerful electric field, a narrow beam is selected and this enters the field of an electromagnetic analyser. The ions sweep round a semi-circular curved path, the radius of curvature of which is related to the magnitude of the accelerating voltage ($\propto \sqrt{E}$) arid to the strength of the magnetic field (($\propto 1/B$)).

The lighter ions are deflected most, and the heaviest, least. Paths are altered so as to bring successive beams through the slit of the collector. On neutralization of the positive ions, suitable electrical amplification produces a signal which can be recorded. Suppose the accelerating field, E, acts on positive ions, mass m, to produce a velocity v and hence a kinetic energy of $\frac{1}{2}mv^2$ for an ion assumed to be of single charge, e.

$$Ee = \frac{1}{2}\ mv^2$$

Ions, equivalent to an electric current of *ev* arriving at the collector have experienced a magnetic field B which caused them to move in a curve of radius r.

$$Bev = \frac{mv^2}{r} \text{ i.e. } Be = \frac{mv}{r}$$

The radius is fixed during manufacture of the instrument, i.e. by elimination of v

$$\frac{m}{e} = \frac{B^2r^2}{2E}$$

Thus ions may be selectively collected according to their *m/e* ratio by altering E or B or both E and B. The isotopic composition of the gas neon could be investigated using the simple mass spectrometer. Neon has three stable isotopes of masses, 20, 21 and 22, of relative abundance 90.92, 0.257 and 8.82 respectively.

On the chart record (Fig. 3), the height of each peak is proportional to the rate of collection of ions. If a photographic plate is used a series of lines is recorded, their positions depending on the radius of curvature of the pathway. By use of a known standard substance in the spectrometer, the recorded mass spectrum may be calibrated and ions identified. Usually the spectrum is drawn as a stick *diagram* with the most abundant species given a height of 100 units (the *base peak)* and each successive peak height proportional to species abundance.

Stick diagrams are used during the identification of ions formed during the breakdown of molecules of an organic compound whose structure is under investigation by mass spectrometry. A simple illustration shows the fragmentation pattern of methane, the mass spectrum in Fig. 1 not being drawn to scale. The stick diagram for

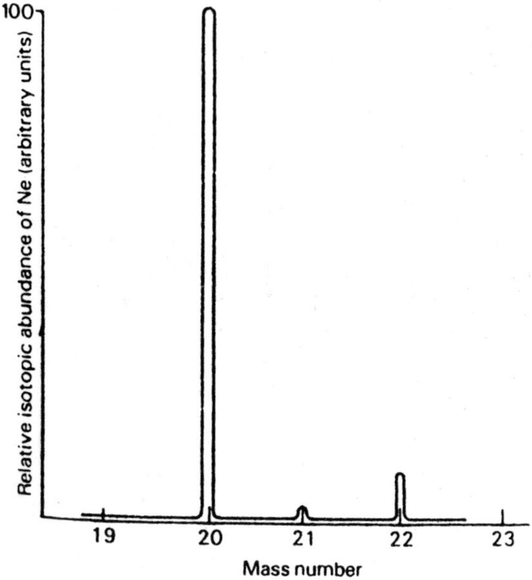

Fig. 1: Diagrammatic representation by mass spectrum of neon recorded by pen on chart.

tetrachloromethane (carbon tetrachloride, $CC1_4$) is much more complicated. Various peaks for $^{12}CC1_4^+$, $^{12}CC1_3^+$, $^{12}CCI_2^+$, $^{12}CCI^+$, Cl^+, $^{12}C^+$ might be expected but each of the species containing chlorine shows multiple peaks due to the isotopes ^{35}CI and ^{37}CI. Thus, there are four peaks corresponding to CCI_3^+

117	$^{35}CI\ ^{35}CI\ ^{35}CI\ ^{12}C^+$
119	$^{35}CI\ ^{35}CI\ ^{37}CI\ ^{12}C^+$
121	$^{35}CI\ ^{37}CI\ ^{37}CI\ ^{12}C^+$
123	$^{37}CI\ ^{37}CI\ ^{37}CI\ ^{12}C^+$

The relative abundance of the isotopes of chlorine may be calculated from the height of the peaks.

Nuclear reactions

We are now in a position to examine the phenomena of

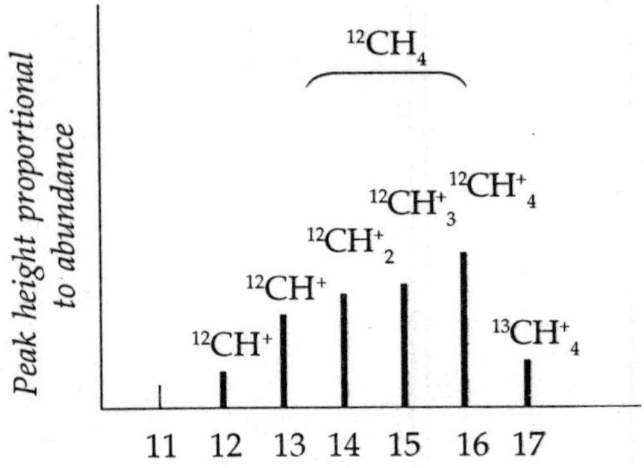

Fig. 2: Stick diagram for CH_4 showing ^{12}C species and a trace of $^{13}CH_4$ not to scale.

radioactivity further. A naturally occurring or generally radioactive isotope may be placed in one of three series of radioactive elements. In the thorium series, in which all mass numbers of elements are multiples of 4 (i.e. mass numbers = $4x$), after ten changes thorium, $^{232}_{90}Th$, forms an isotope of lead, $^{208}_{82}Pb$. The changes involve the formation and disintegration of other elements, including another isotope of thorium:

$$^{228}_{88}Ra \rightarrow ^{228}_{89}Ac \rightarrow ^{228}_{90}Th \rightarrow ^{224}_{88}Ra \rightarrow ^{220}_{86}Rn$$

$$\rightarrow ^{216}_{84}Po \rightarrow ^{212}_{82}Pb \rightarrow ^{216}_{85}At) \rightarrow ^{212}_{83}Bi$$

$$\rightarrow ^{212}_{84}Po \rightarrow (or\ ^{208}_{81}TI) \rightarrow ^{208}_{82}Pb$$

In the uranium series (mass numbers - $4x + 2$), uranium $^{238}_{92}U$, forms a different isotope of lead, $^{235}_{92}Pb$, after fourteen reactions. The first member of the actinium series (mass numbers = $4x + 3$) is now known to be an isotope of uranium, $^{235}_{92}U$, which forms yet another isotope of lead , $^{207}_{82}Pb$, after eleven changes. Another series, the neptuniurn series (mass numbers = $4x + 1$), has been discovered for certain artificially prepared radio active elements; it ends with an isotope of bismuth, $^{209}_{83}Bi$. At each stage in a radioactive series, an unstable atomic nucleus loses energy by the emission of an α-particle (helium nucleus, $^{4}_{2}He$) and/or β^-particle (electron, $^{0}_{-1}e$) and probably γ-radiation. Thus, the emission of a (negative) β-particle (electron) requires the conversion of one neutron into one proton in the nucleus, the atomic number rising by one unit.

$$^{1}_{0}n \rightarrow {}^{1}_{1}p + {}^{0}_{-1}e$$

Emission of energy as γ-radiation leaves the nucleus in a more stable state, with less energy but with the same nuclear charge. Loss of an α-particle (helium nucleus) means that the nucleus loses two protons and two neutrons. Rutherford bombarded nitrogen with α-particles, hoping to cause transmutation of nitrogen.

He was able to demonstrate that protons were formed. Blackett showed that the collision of a fast-moving helium nucleus with a nitrogen atom yields a proton and an atom of oxygen of mass number 17 as distinct from the more usual isotope of mass number 16:

$$^{14}_{7}N + {}^{4}_{2}He \rightarrow {}^{1}_{1}H + {}^{17}_{8}O$$

A more compact notation, in which a represents $^{4}_{2}He$ and p, $^{1}_{1}H$ expresses this change as

$$^{14}_{7}N\ (\alpha, p)\ {}^{17}_{8}O$$

or generally

initial nuclide (incoming particle(s), outgoing particle(s) or quanta) final nuclide

If *n* represents the neutron, and γ the photon (of radiation), write full equations for the following changes and then read on

$$^{59}_{27}\text{Co } (n, \gamma) \ ^{60}_{27}\text{Co}$$

$$^{23}_{11}\text{Na } (\gamma, 3n) \ ^{23}_{11}\text{Na}$$

$$^{27}_{13}\text{AI } (\alpha, p) \ ^{30}_{14}\text{Si}$$

In nuclear reactions the total (mass + energy) of the nuclei is conserved. In addition the total kinetic energy of the moving particles is unaltered. The upper figures in the equations show the mass numbers, atomic masses to the nearest integer, and the lower figures, the atomic numbers or charges.

The sum of the mass numbers on one side of the equation must equal the sum of mass numbers on the other side. Similarly, the atomic (proton) numbers or charges must balance. Energy-mass changes ($E = mc^2$) have too small an influence on the mass to affect the equation although they may be calculated. Returning to natural radio-activity, a series of radioactive changes may be summarized in an equation.

The decay of uranium-238, $^{238}_{92}U$, involving fourteen changes, has been mentioned:

$$^{238}_{92}\text{U} \rightarrow \ ^{206}_{82}\text{Pb} + 8 \ ^{4}_{2}\text{He} + 6 \ ^{0}_{-1}e$$

The decay of radium, found in the actinium series, yields a different isotope of lead after six changes:

$$^{223}_{88}\text{Ra} \rightarrow \ ^{207}_{82}\text{Pb} \ ^{4}_{2}\text{He} + 2 \ ^{0}_{-1}\text{e}$$

In both cases helium nuclei (α-particles) and electrons (β^- particles) are emitted. Most changes are accompanied by γ-radiation.

Nuclear fusion

Nuclear reactions involving the aggregation of light nuclei to form heavier nuclei are called fusion reactions. Energy is evolved in these processes and it is believed that the energy of the sun is provided by fusion of hydrogen nuclei:

$$^{1}_{1}H + ^{1}_{1}H \rightarrow ^{2}_{1}H + ^{0}_{+1}e + \text{light}$$

$$^{2}_{1}H + ^{1}_{1}H \rightarrow ^{3}_{2}He + \gamma$$

$$^{3}_{2}He + ^{3}_{2}He \rightarrow ^{4}_{2}He + 2\ ^{1}_{1}H$$

Helium is the product.

Nuclear power

Calder Hall (and the nuclear/radioactive processing plant at Windscale) near Sellafield in Cumbria opened as the first industrial scale nuclear electricity power station in the world. There was early optimism over prospects for generating abundant, comparatively cheap electrical power as a peaceful application of the nuclear fission process, and as it turned out, wild optimism over the illusive fusion phenomenon.

It would be idle to pretend that military applications were not associated with the research and development work on the chemical treatment of nuclear materials. The engineering problems associated with the construction of nuclear reactors were formidable and pioneering considering that malfunctions needed to be anticipated and health hazards avoided when nobody could for see entirely what would happen as reactor materials aged under continuous running conditions. To the original construction costs must be added those of decommissioning such stations after their 20-30 year economic life span and management of the highly radioactive materials remaining. The Magnox stations come to mind in this context.

The United Kingdom has encountered commercial disappointments over the running of the more advanced gas cooled reactors since their performance has rarely reached expectations. In addition, recent calculations of the 'time cost' of generating electricity in this way has made such stations seem uneconomic in comparative, terms. Much depends, however, on what is meant by time cost'. The time cost of electricity generated by combustion of fossil fuels (coal, oil) in an ideal world would include a factor to cover environmental protection from the sulphurous products causing 'acid rain' and respiratory problems in those people prone to such weakness. As an alternative to the (wasteful) use of fossil fuels, nuclear power has many attractions especially where supplies of such fuels are not abundant.

Most of the electricity generated in France comes from nuclear sources and the commercial price, when supplies are exported via cables to adjacent countries (including England at peak times), is very competitive. Medium sized nuclear power stations with their characteristic hemispherical reactor shape can be seen in West Germany, and there is much to be said for such smaller stations. There are alternative ways of generating electrical power which are becoming popular. Hydroelectric schemes have been developed for many years and at last the possibility of harnessing the energy of wave motion and tides is being explored.

Along the North Sea coast, in Holland, and in Schleswig-Holstein and Denmark especially, tall, elegant two or three bladed 'windmills' (2-4 times the height of traditional windmills) dot the landscape, their propellor-like blades twirling in the breeze.

Hazards of nuclear power and processing

Design faults, inadequate engineering techniques disease and the so-called human factor have resulted in various nuclear power station incidents in the USA and in Europe, causing public alarm. When coupled with a growing cynicism about how forthcoming governments are over such incidents (the Macmillan government was less than candid over an emission of radioactivity from

Windscale) it is hardly surprising that people can develop an antipathy towards the nuclear power industry. An international catastrophe such as the explosion at the Russian nuclear power station at Chernobyl caused serious loss of life and lowered life-expectancy in the area. The effects were also experienced in other countries, as seen in the serious blow to the livelihoods of sheep farmers as far away as Cumbria when their meat was barred from the market.

The discovery that the Greifswald nuclear power station, near to the military airfield at Peenermumde (with an inherent risk of aircraft crashes) in East Germany, was without a concrete shield over the reactor led to its closure. The closure followed inspections by West German experts during preliminary unification meetings and makes one wonder about the quality of design and construction there and elsewhere. Numerous incidents at reactors (many of little significance) and agitation over the location of radioactive processing plants have cast a political shadow which could blight an industry which should have enjoyed an assured future. The biological damage caused by nuclear irradiation is insidious and the long term effects of exposure, even to seemingly safe radiation levels, tragic.

There is, of course, the other side of the coin-the beneficial effects of radiation treatment for cancer. However, it remains that the cost of Chernobyl still has to be paid for by some with their lives. In the United Kingdom, a television programme, *Windscale: The Nuclear Laundry* first suggested that there were unusually high incidents of blood cancer leukaemia around Sellafield, the former Windscale plant.

The *Black Report* concluded that the increased incidence of leukaemia was real but that there was nothing in the environment (including radioactive discharges) to account for them. The Committee on Medical Aspects of Radiation was set tip and found six cases of leukaemia in children where only one would have been expected.

The normal chance of contracting the disease is 1 in 2000. In the previous year *The British Medical Journal* had highlighted other cases. At nuclear plants at Aldermaston and Burghfield there were 41 cases of leukaemia among children under 14 years of age where only 28 would have been expected. Other data were collected but no investigation could explain the cause, even taking the local carcinogenic bracken into account. The *Garden Report* suggested a genetic link between parental occupational exposure and the appearance of leukaemia, suggesting that a father experiencing higher than normal radiation levels may carry genetically damaged sperm cells.

The report stated, 'It is something we suggest is biologically possible, rather than proven'. Sellafield is situated in a bleak, beautiful region that has limited economic prospects. British Nuclear Fuels Ltd (BNFL), a major employer, processes some of the UK nuclear power waste there. Together with its workers it has to face health hazard problems which exceed the range and depth of those faced by other industrial companies, by far. Finally, it must be remembered that we all experience background radiation, the intensity of which depends on where we live and our occupation. We are all at risk to various degrees, however slight this may be.

The detection of radioactivity

The image produced on a photographic plate by a crystal of a uranium salt led Becquerel to the discovery of radioactivity. The plate was protected from the action of visible light. The action of radiation on a photographic emulsion remains one of the simplest tests for the detection of radiation from radioactive sources and X-ray equipment. In radiation laboratories and industrial plants where there is the risk of exposure to harmful radiation, workers carry suitably prepared pieces of film which are developed to check that exposure has not been excessive. There is a necessary time lag for development and another principle is used for the immediate detection of radiation. Ionization of a gas is caused by charged particles (α, β^-, β^+), and by particles (β^-) resulting from the interaction of X- and y-radiation with the gas.

Collision of a charged particle with a gas molecule may result in the ejection of an electron, leaving behind a positively charged ion. The Wilson cloud chamber is used in laboratory experiments, and the Geiger-Muller counter ('Geiger counter') may be used in the laboratory and as a portable detector. In the Wilson cloud chamber, dust-free air is saturated with water vapour and cooled suddenly by adiabatic expansion, becoming supersaturated. Condensation occurs on the ions produced and the track of each sub-atomic particle is revealed as a trail of fine droplets.

With suitable illumination, the track may be photographed. In the Geiger-Miffler counter, a small cylindrical ionization chamber contains an ionizable gas (air or argon) at low pressure, the wall acting as the negative electrode. A central axial wire acts as the positive electrode and is insulated from the wall of the chamber. A potential difference in the range 1000-2000 V is maintained between the electrodes. This must be insufficient to cause electrical discharge but enough to drive all ions produced to the electrodes. When a charged particle enters the gas, ionization occurs.

The ions produced are accelerated sufficiently to cause further ionization. Each particle causes a burst of ionization. By suitable amplification an electric current may be measured, the bursts of ionization counted in a counting device or made audible through a loudspeaker. The circuit is arranged so that the discharge lowers the potential between the electrodes and ionization ceases until the next particle enters the chamber.

Radioactive isotopes

The course of a chemical process may be followed by substituting a radioactive isotope of an element for the normal non-radioactive form. Chemically they are identical and their isotopic ratio remains essentially unchanged. By determining the intensity of the radioactivity due to the isotope, the concentration of the element may be calculated, even in very small concentrations. This is seen at once to be especially useful in biological systems". Photosynthesis has been studied using radioactive phosphorus in the phosphate supplied. Of special interest is the estimation of the date at which animals or vegetation died by measuring the intensity of radioactivity due to the carbon-14' isotope. During their lifetime plants and trees take up small quantities of $^{14}_{6}C$ as carbon dioxide, the quantity reaching an equilibrium value.

Plants are eaten by animals which also acquire carbon-14. At death, the intake is stopped and radioactive decay continues. The half-life period is 5720 years. By determining their radioactivity the age of various carbon products has been measured, covering a period of 1000-30 000 years, by Libby.

Uses of typical elements

The production of the commerically more important elements and their uses are presented in this chapter, group by group.

Lithium, sodium and potassium

Lithium

Occurrence

Although lithium is roughly twice as abundant as lead, economically worth while deposits of lithium compounds are comparatively rare.

The ores are complex. The principal sources are spodumene, lithium aluminium silicate,

$$LiAl(SiO_3)_2$$

lepidolite, a 'lithium mica'

$$K(Li,Al)_3(Al,Si)_4O_{10}(OH,F)_2,$$

petalite, another lithium aluminium silicate,

$$(Li,Na)AlSi_4O_{10},$$

and amblygonite, a fluorophosphate,

$$(Li,Na)AlPO_4(F,OH).$$

All are mined in Zimbabwe and Namibia. The first mineral mentioned also occurs in the USA.

Production

Lithium sulphate is made by roasting petalite, lithium aluminium silicate, with sulphuric acid. The carbonate is then precipitated under carefully controlled conditions and converted into the hydroxide in the Gossage manner, using slaked lime:

$$Li_2CO_3 + Ca(OH)_2 \rightleftharpoons CaCO_3 + 2LiOH$$

The least soluble compound, calcium carbonate, is precipitated and removed. On evaporation, the monohydrate of lithium hydroxide separates on cooling to 40°C. Carbon dioxide must obviously be excluded. The product is converted into lithium chloride by the action of hydrochloric acid. Electrolysis of a fused mixture of lithium chloride (60%) and potassium chloride (40%) at

475-500°C with a steel cathode, separated from graphite anodes by a wire screen, yields lithium at the cathode.

Uses

In addition to the uses of lithium compounds in organic synthesis, lithium metal is employed as a de-gasifier in metallurgy and by the German railways as a minor constituent (0.04% Li) with lead in Bahnmetall for anti-friction bearing alloy. Missile fuels have been developed from the lithium oxosalts, nitrate and perchlorate. Lithium hydride, because of its reactivity and low relative molecular mass (molar mass), is an extremely convenient portable source of hydrogen, generated by reaction with water, used for air-sea rescue work and in observation balloons. The hydrogen bomb and thermonuclear processes probably depend on lithium deuteride in conjunction with plutonium and uranium.

Beryllium, magnesium and calcium

Beryllium

Occurrence

Beryllium is listed as 32nd in the order of estimated abundance of elements in the Earth's crust. The principal mineral is beryl, beryllium aluminium silicate, $2BeO.Al_2O_3.6SiO_2$. This is bluish-green in colour. The deep green translucent form is the precious stone, emerald. The ore is worked in Brazil chiefly but also in Argentina, Zimbabwe, South Africa and India.

Production

Beryl crystals, having about the same density as the accompanying silica, are hand-sorted. The beryllium content is roughly 2½% with iron as the main impurity. The first stage is the isolation of beryllium hydroxide. The ground ore is heated with lime to 1500°C to react with the alumina and silica present.

Addition of concentrated sulphuric acid precipitates silica and calcium sulphate, giving a liquor from which ammonia alum is crystallized on addition of ammonium sulphate. Iron is precipitated as hydrated iron(III) oxide after oxidation of the iron (II) to the iron(III) state by sodium peroxide and adjusting the pH (= 4.0) with chalk.

Beryllium hydroxide is precipitated by addition of ammonia, removed and dried at 100°C. The hydroxide is bricketted with charcoal using wood tar as a binder, baked to the oxide and chlorinated at 700-800°C in an electrically heated furnace when beryllium chloride volatilizes away with the impurities: chlorides of aluminium, silicon and iron(III);

$$BeO + C + Cl_2 \rightarrow BeCl_2 + CO$$

Purification is effected by fractional distillation in an atmosphere of hydrogen, which reduces any residual iron(III) chloride to the non-volatile iron(II) chloride.

Beryllium chloride, mixed with an equal weight of salt, is electrolysed in nickel cells heated externally to 350°C. Chlorine is evolved at the graphite anode and beryllium collects on the walls and bottom of the cell, which acts as the cathode. The metal is flaked off and sintered at 1400°C in hydrogen to give metal of 99.0-99.5% purity.

Uses

Beryllium copper (up to 40% Be), an alloy of tensile strength greater than that of other non-ferrous alloys and comparable to that of alloy steel, was the, chief application until recently. The search for metals with the special characteristics needed for atomic reactor equipment has stimulated research into beryllium production, but it is difficult to work and when finely divided, is toxic. It has been used as a sheathing metal for atomic fuel cans to contain the radioactive metal and fission products, and in advanced gas-cooled reactors designed to work up to 600°C, as it absorbs relatively few neutrons, is chemically inert to both fuel and coolant gas and conducts heat well. Beryllium is used in X-ray 'windows', and in high-speed aircraft, having superior hardness, elasticity and strength to magnesium and aluminium but with a density slightly more than that of the former and only two-thirds that of the latter. Beryllium has been used in the construction of man-made Earth satellites and in rocket guidance equipment.

Magnesium

Occurrence

Magnesium forms an estimated 2.2% by weight of the Earth's crust. It is the eighth most abundant element and the sixth most abundant metal. Solid deposits are widespread, especially dolomite, $MgCO_3 . CaCO_3$. Magnesite, $MgCO_3$, occurs in North America, Europe, India and the USSR. Brucite, $MgO. H_2O$, is found in Canada. Mineral deposits formed by the evaporation of inland seas, such as those at Stassfurt in Germany (the former DDR), contain double salts, e.g. carnallite, $MgCl_2 . KCl. 6H_2O$ and kainite, $MgSO_4 . KCl. 3H_2O$, and kieserite, $MgSO_4 . H_2O$.

However, sea water is the main commercial source of magnesium. It contains 0.13% Mg^{2+} by weight. Evaporation of sea water to roughly one-sixth of the original volume results in the separation of much common salt and a liquid remains which is rich in magnesium salts. These impart a bitter taste. The solution is called a 'bittern'. Because of their strength, lightness and ease of working, magnesium alloys are of increasing importance, but comparatively few countries can produce magnesium economically. In the UK, magnesium production has fluctuated: it was economical during the period. German competition proved too strong, in the First World War and again during the period. Production started again.

The cheapest magnesium is formed by electrolysis of the fused chloride, in Norway and at Freeport, Texas, USA. Low cost of raw materials, including sea water and an abundant electricity supply, favours Norway, while at Freeport, sea water and the purest lime in inexhaustible supply with natural salt deposits, natural gas as a source of heat and electrical energy, provide ideal conditions. Natural advantages of mineral wealth and electric power make Canada and Italy important of magnesium by chemical reduction.

Production by electrolysis

The electrolysis of fused magnesium chloride yields magnesium and chlorine. Originally carnallite was used but the process was inefficient due to the hydroxide ions present in the basic salt. Hydrogen was liberated, thereby lowering the yield of magnesium, and wear occurred at the anodes due to the liberation of oxygen. Carnallite was replaced by synthetic magnesium chloride. Two methods of obtaining magnesium chloride will be described but the principal source is sea water.

In the Electron process, magnesite is calcined at 500°C to give magnesium oxide,

$$MgCO_3 \rightarrow MgO + CO_2$$

The product is bricketted with fine coal using magnesium chloride as binder, coked in rotary kilns and chlorinated above the melting-point of magnesium chloride:

$$MgO + Cl_2 + C \rightarrow MgCl_2 + CO$$

$$2MgO + 2Cl_2 + C \rightarrow 2MgCl_2 + CO_2$$

The anhydrous fused magnesium chloride is transferred to the electrolysis pots as required.

In the Dow process, magnesium is extracted from sea water and from the residual liquors remaining after the recovery of chlorine, bromine, sodium and calcium from the brine wells in Michigan, USA. Sea water is run into settling tanks and mixed with milk of lime. This is formed by slaking the quicklime produced when oyster shells, composed of calcium carbonate and dredged from the ocean floor, are roasted. Magnesium hydroxide is precipitated.

Hydrochloric acid, formed by interaction of chlorine and natural gas, is added. Magnesium chloride is formed. The solution is evaporated to dry flakes, hydrolysis being avoided by leaving the salt partially hydrated. Electrolysis is done in steel cells lined with fire brick and heated externally.

The temperature is 670-730°C and the melt contains sodium chloride (50-60%) and calcium chloride (15%) as fluxes and to increase the density. Magnesium, released at steel cathodes, floats to the surface where it is protected from oxidation by an atmosphere of coal gas. Chlorine is liberated at graphite anodes and ducted away by a pipe surrounding the electrodes to avoid the interaction of the products.

The electrode processes are:

$$Mg^{2+} + 2e^- \rightarrow Mg \quad 2Cl^- - 2e^- \rightarrow Cl_2$$

The current is 30000 - 50000 A at 6.6 V potential difference and the product, 99.5-99.9% magnesium, is ladled into moulds. Chlorine is re-cycled.

Production by chemical reduction

Magnesium oxide is very stable but processes have been developed to effect reduction. These necessarily require very high temperatures and, unlike electrolysis, have the disadvantage of being non-continuous.

The Hansgirg process requires reduction of calcined magnesite with coke,

$$MgCO_3 \rightarrow MgO + CO_2$$

$$MgO + C \rightleftharpoons Mg + CO$$

The reaction is brought about at 2000°C in an electric arc furnace with the quenching of the vaporized products to 200°C by hydrogen gas, or natural gas (where available), which is subsequently burnt. This avoids the reduction of carbon monoxide by magnesium. The metal, in powder form, is redistilled at 800°C in a vacuum and the crystalline solid product collected.

Uses

Contrary to what might be at first thought, magnesium is not dangerously flammable. The pure metal is not strong enough for structural use by itself so a range of light alloys has been developed. An alloy is a substance with metallic physical properties formed from a metal by the addition of other elements, usually metals but sometimes metalloids or non-metals. Desirable modifications of magnesium can be brought about by the addition of aluminium, manganese, zinc and zirconium.

Magnesium is very reactive, and in addition to its being placed high in the electrochemical series there is no hydrogen overpotential. It is used in cathodic protection of other metals, including iron and steel structures: buried pipe-lines, buried structures of all sorts, ships' hulls and the insides of catalytic crackers and distillation plants. In certain circumstances, magnesium alloys require protection.

Thus, at joints with other metallic surfaces where moisture cannot be excluded, special jointing compounds are used. Special chemical treatment with a chromate solution induces surface passivity, which can also be produced by electrolytic anodizing. However, epoxy resin paints provide the best protection. Magnesium alloys are welded in an atmosphere of argon by the tungsten arc process. Alloys are used in manufacturing atomic fuel cans for gas-cooled reactors. Magnesium is used in the reduction of titanium(IV) chloride to the metal in pyrotechnics and photographic flash work.

Calcium

After aluminium and iron, calcium is the commonest metal found combined in the Earth's crust. It is essential to both animals and plants. Calcium phosphate is the chief constituent of bones. Calcium carbonate occurs widely as limestone, chalk, marble, and also as calcite, Iceland spar, aragonite, etc., or with magnesium carbonate as dolomite, $CaCO_3 . MgCO_3$.

Calcium sulphate is mined as anhydrite, $CaSO_4$, and as gypsum or alabaster, $CaSO_4 . 2H_2O$. Calcium silicates are found in rocks. Calcium is produced by the electrolysis of fused calcium chloride. It is used in the laboratory as a reducing agent, for removing traces of hydrogen and nitrogen from gases and as a deoxidizer in some metallurgical operations.

Boron and aluminium

Boron

Comparatively little elemental boron is produced, although boron compounds are used widely. The element's principal source lies in southern California, USA, where deposits of sodium borates, kernite, $Na_2B_4O_7 . 4H_2O$, and (tincal) borax, $Na_2B_4O_7 . 10H_2O$, are mined from the Mojave Desert. The borax industry is described later. A black or brownish-black amorphous boron is produced by reduction of the oxide with sodium or magnesium, or fluoroborate with potassium; it is very difficult to purify:

$$B_2O_3 + 6Na \rightarrow 2B + 3Na_2O$$

$$KBF_4 + 3K \rightarrow 4KF + B$$

Almost pure crystalline boron is made by the reduction of boron bromide vapour with hydrogen at 1000-1300°C on a tantalum filament just below atmospheric pressure:

$$2BBr_3 + 3H_2 \rightarrow 6HBr + 2B$$

Uses

Among the elements which are readily available, boron is second only to cadmium as a neutron absorber. The element and boron carbide are used in the shielding of atomic piles, screening operators, and in the control rods used for regulating the chain reaction.

Having a low density (2.3 g cm^{-3}), boron has strong claims for use in mobile nuclear reactor units e.g. in submarines and surface ships. High-purity boron is used as a crystalline semiconductor material and has the advantage of functioning at higher temperatures than germanium and silicon, the limits for which are approximately 100°C and 300°C, respectively. Boron is an essential minor element for the healthy growth of plants.

Aluminium

Aluminium forms an estimated 7.5% by weight of the Earth's crust. It is the most abundant metal, occurring as silicates in clay, granite, shale and slate, and as the hydrated oxide in the bauxites. The latter are formed by the weathering of clays in warm regions of abundant, intermittent rain falling on land of low relief so that the rain soaks into the deposits, washing out silica and leaving aluminium oxide. The hydrated aluminium oxide of bauxite may occur in a number of forms: gibbsite, Al_2O_3. $3H_2O$, as böhmite and diaspore, Al_2O_3. H_2O. They range in colour from creamy-white to reddish-brown, having an aluminium content of 50-65%.

The main impurities are oxides of titanium, silicon and iron. White bauxite has 1-4% silica and no iron while red bauxites contain up to 27% red iron(III) oxide. Bauxite occurs in widespread but localized deposits which are 3-30 m deep. At present, the extraction of aluminium from clay is uneconomic and it is made solely from bauxite. The degree of hydration varies from Al_2O_3.

H_2O in the French deposits to Al_2O_3. $3H_2O$ in Guyana. Deposits are exploited in south-eastern Europe, in the USSR, China, India and southeastern Asia, Australia, the West Indies, Brazil and southern USA. Bauxite is quarried by drilling and blasting. Coarse impurities are removed and before transportation it is washed, dried and graded.

The only known source of the other aluminium mineral, cryolite, is at Invigtut in Greenland from where it is exported to the UK. Much sodium aluminium fluoride, or synthetic cryolite, is made industrially for the extraction process.

Preliminary treatment of bauxite

Bauxite is crushed and digested with caustic soda solution under pressure. The resulting liquor is diluted and an insoluble residue containing titania, titanium(IV) oxide, sodium aluminium silicate and iron(III) oxide settles. The supernatant liquid containing sodium aluminate is collected and filtered.

A large excess of crystalline aluminium oxide trihydrate, Al_2O_3. $3H_2O$, is stirred into the solution. This acts as a seeding agent and crystalline trihydrate is deposited instead of the more usual gelatinous precipitate. It is filtered, washed and calcined at about 1300°C to give anhydrous alumina. Although the chemical changes are no doubt complex, the reactions may be represented:

$$Al_2O_3.\ 3H_2O + 2NaOH \underset{\text{decomposition}}{\overset{\text{extraction}}{\rightleftharpoons}} \underset{\substack{\text{sodium}\\\text{aluminate}}}{2NaAlO_2} + 4H_2O$$

in which sodium aluminate solution is only stable with an excess of free alkali present. The left to right reaction is extraction, the reverse, decomposition.

An alternative formulation of the aluminate anion as a hydroxo-aluminium complex:

$$Al_2O_3.\ 3H_2O + 2NaOH \underset{decomposition}{\overset{extraction}{\rightleftharpoons}} 2NaAl(OH)_4$$

$$Al_2O_3.\ 3H_2O + 2OH^- \rightleftharpoons 2Al(OH)_4^-$$

On calcination,

$$Al_2O_3.\ 3H_2O \rightarrow Al_2O_3 + 3H_2O$$

The smelting process

Aluminium is formed by electrolysis of alumina dissolved in molten cryolite, between carbon electrodes. Calcium fluoride (fluorspar) is added to lower the melting-point. An outline theory gives the following dissociations:

$$2AlF_3 \rightleftharpoons Al^{3+} + AlF_6^{3-}$$

$$Al_2O_3 \rightleftharpoons AlO_2^- + AlO^+ \quad AlO \rightleftharpoons Al^{3+}\ O^{2-}$$

Aluminium is preferentially discharged at the cathode and oxygen at the anodes:

$$Al^{3+} + 3e^- \rightarrow Al$$

$$2AlO_2^- \rightarrow Al_2O_3 + \tfrac{1}{2}O_2 + 2e^-$$

The carbon anodes burn to carbon dioxide,

$$C + O_2 \rightarrow CO_2$$

This is the primary by-product but some aluminium is dispersed as fog in the electrolyte, reducing dioxide to carbon monoxide. Exit gas contains about 30% carbon monoxide.

Undoubtedly such complexes as $Al_2OF_6^{2-}$ and $Al_2O_2F_4^{2-}$ are present, depending on the aluminium concentrations. The overall anode reaction seems to be

$$Al_2O_2F_4^{2-} + 2AlF_6^{3-} + C \rightarrow 4AlF_4^- + CO_2 + 4e^-$$

and at the cathode,

$$AlF_6^{3-} \rightleftharpoons 3e^- \rightarrow Al + 6F^-$$

the hexafluoroaluminate ion being in equilibrium with the tetrafluoroaluminate ion

$$AlF_6^{3-} \rightleftharpoons AlF_4^- + 2F^-$$

The anode process may be seen as a smelting in which sodium aluminate solution is only stable process superimposed on an electrolysis. Any with an excess of free alkali present. The left to right calcium (from the flux) deposited is volatilized reaction is extraction, the reverse, decomposition away.

An overall cell reaction may be computed:

$$3Al_2O_2F_4^{2-} + 10AlF_6^{3-} + 3C \rightarrow 12AlF_4^- + 3CO_2 + 4Al + 24F^-$$

A current of about 40000 A passes in series through about 100 cells. The voltage drop across each cell is about 5 V. Aluminium accumulates in a pool *c*. 100 mm deep on the floor of the cell, while the anodes, suspended *c*.50-75 mm above the surface of the aluminium, burn to gaseous products, which bubble away. More alumina is added to replace that used and the crust is plugged. As the concentration of alumina falls, fluorine is evolved and the voltage required for electrolysis rises to 30-50 V. This happens every 8 hours.

About 275 kg of aluminium is formed daily in each cell and every two days it is either siphoned out by suction or tapped, depending on the design of cell used. The fluid metal is transferred to an oil or gas-fired reverberatory holding furnace at 900°C. From this furnace aluminium is cast into ingots or rolling blocks. Because the density of the metal (2.3 g cm^{-3}) is not very different from that of its electrolyte (2.1 g cm^{-3}), some aluminium is dispersed as a fog and is oxidized, causing a loss in efficiency.

Although the metal is liberated in accordance with

Faraday's Laws, the process is only about 85% efficient. In practice about 6.8 kg, instead of 8.0 kg aluminium, is liberated by a current of 1000 A in one (24 h) day. This charge at 5 V potential corresponds to an energy expenditure of 120 kW h, so that every tonne (1000 kg) of aluminium requires about 18000 kW h of electrical energy†. Cheap hydro-electric power is essential. Aluminium is manufactured in mountainous regions with a generous rainfall, including the western Highlands of Scotland, Norway, Switzerland, parts of France and Italy, and the Canadian Rocky Mountains.

Of the widely used metals, aluminium is by far the most expensive to produce in energy terms, which has given an impetus to the recycling of the metal, the collection, melting and reworking of scrap metal. Other methods of extraction may be developed, that involving the chloride as intermediary seeming attractive. The formation of 1000 kg of aluminium requires about 2000 kg of alumina and results in the combustion of about 650 kg of carbon anode. To reduce costs, Soderberg electrodes were introduced in which a paste of coke and pitch-tar binder is confined in a casing and bakes to a hard mass as it nears the electrolyte.

Ash from the anodes introduces impurities into the aluminium. Anthracite is used in the preparation of cell linings. The aluminium is up to 99.8% pure, with traces of iron and silica. Commercial aluminium is never less than 99% pure. Further purification need not concern us here.

Uses

Pure aluminium is a soft white metal of low tensile strength, but alloying and heat treatment brings the strength near to that of mild steel. Aluminium alloys are used for aircraft frames, car engines and components, bridges, building structures, superstructures on ships, scaffolding and ladders, where lightness is important.

Aluminium has a density of only a third that of steel. High electrical conductance and low density give aluminium a great advantage over copper when used for overhead electrical transmission lines.

Steel-cored aluminium cables are used in the National Grid. Although aluminium has only two-thirds the electrical conductance of copper, the density of one-third that of copper gives it twice the electrical conductance on a weight-for-weight comparison. Aluminium is characterized by high resistance to corrosion, due to a thin film of oxide, which may be increased in thickness by anodizing.

High thermal conductance and resistance to corrosion lead to the application of aluminium in kitchen utensils and the food industry. The reflecting power of the metal, coupled with the lightness and resistance to corrosion of aluminium, bring the application of aluminium alloys as a roofing material, giving thermal insulation by reflecting the heat of the sun. Similarly it is used in the linings of cold-weather clothing (e.g. anoraks) and in survival bags for emergency use on mountains.

The appearance of the metal is an asset and the surface may be left unpainted in, many of its applications, thereby giving an economic advantage over other materials, which need protection, and even an extra economy of lower weight (no paint!) for London Underground trains.

Carbon, silicon, germanium, tin and lead

Graphite (carbon)

Occurrence and production

Deposits of graphite or plumbago are found in Ceylon and Central America. It was formerly mined in Cumberland, being used to make 'black lead' for pencils. Another allotrope of the element is mined in the form of diamonds in South Africa and the USSR.

Carbonates and organic compounds are widespread. Graphite is manufactured by the Acheson process, which requires abundant cheap electrical power. Petroleum coke and sand, with coal tar as a binding agent, are maintained at a very high temperature for 24-30 hours in an electric furnace. The alternating current is led through carbon rods which pass through the mass, supported on a refractory hearth. It is thought that silicon carbide is formed and then dissociates to graphite and silicon, which volatilizes away:

$$SiO_2 + 3C \rightarrow SiC + 2CO$$

$$Sic \rightarrow Si + C$$

Uses

Black lead is produced by mixing graphite with moist plastic clay, extruding into threads and baking. Pure graphite is used for electrodes, brushes for electric motors and carbons for dry cells, because of its good electrical conductance. It is an excellent lubricant and may be used as a colloidal suspension in water or in oil.

The suspension in water is stabilized by the presence of tannin. The commercial name 'Dag' is short for deflocculated Acheson Graphite. Carbon fibre has been developed for use in engineering. Rolls Royce aircraft engines are a magnificent example of the product of research and (costly) development of this material, and many of us prize our carbon fibre fishing rods.

Silicon

Occurrence

Silicon is our second most abundant element and is always found combined with the most abundant element, oxygen. It is estimated that silicon forms 27.6% by weight of the Earth's crust. Silicon dioxide, silica, SiO_2, is found comparatively pure as quartz, flint, agate and sand, and is formed by weathering of igneous rocks, which contain silicates. An example of such rock is orthoclase or potash felspar, $K_2O.\ Al_2O_3\ .6SiO_2$. Under the action of rain water containing carbon dioxide, acid water from peat bogs, etc., and those regions where volcanic gases occur, the complex silicates disintegrate to form clays and by further changes, silica. China clay has the composition $Al_2O_3.\ 2SiO_2.\ 2H_2O$.

Uses

Pure silicon has poor mechanical properties and rinds application in alloy form only. Ferrosilicon, usually containing iron and silica in equal amounts, is used for the de-oxidation and de-gasification of steel formed by the Open Hearth and Bessemer processes. It is produced by heating high-grade silica with coke and iron in an electric furnace, It is common practice to add lumps of ferrosilicon to acid open hearth steel just prior to tapping. To avoid reduction of phosphates to phosphorus in the basic process, the addition of crushed ferrosilicon is made to the steel in the ladle after tapping.

Structural steels contain less than 0.3% silicon but other steels contain as much as 0.6% to increase hardness and generally to improve their properties. Electrical steels and certain special steels use much more. Silicon of metallic appearance is made by the charcoal reduction of pure silica and is used in alloying. Silumin, a product of the reduction of china clay, containing up to 12% silicon with aluminium, combines strength with hardness and a low coefficient of thermal expansion, which makes it a useful material for pistons and cylinders in car engines.

The oxide, silica, and silicon carbide are used as refractory materials and the latter under the trade name of Carborundum is also used as an abrasive. Silicon 'chips' are used in micro-electronic circuits. The name has become a household word. The devices are developed from the original invention, based on discoveries using pure germanium, at the Bell Telephone Laboratories in New York. The development of zone-refining and allied techniques enabled very pure silicon to be formed and certain unique aspects of its chemistry made the manufacture of integrated circuits possible, so that silicon has displaced germanium in this area.

The initial impetus came from the need for maximum reliability and minimum weight for the intercontinental ballistic missiles developed for the US Defence Programme. Miniature (or integrated) circuits combining a range of electrical functions may be made all on the surface of a single piece of semiconductor material.

The dimensions of such small components are expressed in micrometres and the development research is intensely competitive. Other materials, e.g. gallium arsenide, have been investigated for use.

Germanium

Occurrence

Germanium is found widespread but in small concentration. Minute quantities occur in silicate rocks and it is associated with certain lead and zinc ores. The dioxide is found in small amounts in certain coal ashes. It is extracted commercially from the flues of gas producer plants, the dust coming from coke ash, and is a by-product after the extraction of zinc. Germanium is used in the electronics industry although the use of silicon is becoming paramount. Germanium minerals include argyrodite, a silver germanium(IV) sulphide, $4Ag_2S.\ GeS_2$, found in Bolivia, and germanite, the sulphide associated with copper and iron(II) sulphide, $7Cu.\ FeS.\ GeS_2$, in Namibia.

Production

By various methods, germanium tetrachloride, germanium(IV) chloride, is formed from compounds associated with zinc and gas producer plants. It is purified by distillation and hydrolysed with water to the dioxide which is washed, dried and reduced in a stream of hydrogen at 600-650°C to yield germanium powder:

$$GeCl_4 + 2H_2O \rightarrow GeO_2 + 4HCl$$

$$GeO_2 + 2H_2 \rightarrow Ge + 2H_2O$$

The powder is melted at 1100°C under a chemically inert gas, such as nitrogen, and cast into ingots which may be further purified by heating to 1000°C under a vacuum with controlled cooling to separate the impurities.

Uses

Germanium crystal diodes and triodes were originally manufactured to replace thermionic valves. They are small and have the advantages of reliability over a long life, with a much greater resistance to mechanical shock. The development of the germanium crystal diode followed the need to detect very high frequencies in early radar work during World War II.

The germanium crystal triode was developed by the Bell Telephone Co., USA, in 1948 and called the 'transistor'. In transistors, crystals of germanium with properties modified by the controlled addition of certain other elements are used to achieve rectification and amplification.

Group III elements—boron, aluminium and gallium—give the Positive (P) Type, and Group V—nitrogen, phosphorus and arsenic—the Negative (N) Type; one type is sandwiched as the 'grid' between two layers of the other type.

Tin

Occurrence

The only important ore of tin is tin(IV) oxide, cassiterite or tin-stone, SnO_2. The principal sources are Malaysia,

Indonesia and Bolivia, while important deposits are worked in Nigeria, Zaire, Thailand and China. Cornwall, a primary source of tin a century ago, now yields comparatively little.

Tin-stone is found in alluvial gravel in the Far East and is recovered by dredging, hydraulic sluicing of gravel banks and pumping from gravel mines. Other deposits yield tin in association with the sulphides of iron, copper, lead, zinc, arsenic, and tungsten as wolframates (tungstates). Tin is also recovered from scrap tin-plate.

Preliminary treatment of cassiterite

The ore is crushed and concentrated by agitation in water, lighter material being washed away. Further concentration is accomplished chemically with the aim of converting the impurities into a soluble or volatile form leaving the tin(IV) oxide unchanged. Roasting enables sulphur, arsenic and antimony to volatilize as oxides, leaving metallic oxides which may be leached out with acids.

Salt added during roasting assists volatilization of impurities as chlorides and subsequent leaching out of what remains. Remaining sulphides may be recovered by froth flotation and magnetic minerals of iron and tungsten by electromagnets. The method adopted in the preliminary treatment depends on the impurities.

The smelting process

The tin concentrate is smelted in a reverberatory furnace with charcoal or anthracite and a limestone flux:

$$SnO_2 + 2C \rightarrow Sn + 2CO$$

If much iron remains, a silica flux is used. Because of the amphoteric nature of tin(IV) oxide, calcium stannate(IV) is formed to some extent with a basic flux and tin silicates with an acidic flux. The slag is resmelted at a higher temperature to achieve further separation of tin.

Liquid tin is run off and cast into slabs for refining.

Refining

Tin smelted from Far Eastern ore requires little refining, the principal impurity being iron. Impure tin is also associated with copper, lead, arsenic, antimony and bismuth when smelted from other sources. The procedure depends on the impurities. Tin has a relatively low melting-point (232°C) and refining by liquefaction exploits this. Impure slabs of metal are heated to melting on the sloping hearth of a small reverberatory furnace: molten tin runs away. The residue is smelted with slag for tin recovery, arsenic and much iron being removed. Preferential oxidation removes other impurities.

Air or steam is blown through strongly heated molten metal when iron and other impurities are oxidized and coagulate to be skimmed away. Similarly, chlorine is used to oxidize lead and sulphur, to remove copper as copper(I) sulphide, which is also skimmed away.

Uses

Tin is a soft, silvery-white, lustrous metal, plastic and malleable, but of low tensile strength and not at all ductile. It is notable for its resistance to chemical action. Tin is subject to an enantiotropic change at 13.2°C with the formation of a grey, powdery and brittle allotrope. Fortunately, the change is subject to a good deal of supercooling. Where the change does occur, the growths of grey tin appear fungoid in character on the white tin surface, justifying the description, tin plague. Tin is used as foil for the wrapping of cheese, chocolate and tobacco, for electrical condensers and cables and in extrusions for collapsible tubes used for cosmetics, pharmaceutical preparations, toothpastes, adhesives and paints, etc., and for pipes conveying beer and carbonated drinks, chemicals and water. Up to 40% of tin produced, appears as coatings on other materials, mainly steel and copper. Tinning is achieved by hot dipping in molten tin, spraying or wiping molten metal on to the hot article or by electrodeposition. Steel tin-plate is used for food cans. Domestic, brewery and dairy utensils may be tinned copper or iron depending on the quality. Tinned copper, tinned brass and tinned cast iron have other uses as well. About 30% of all tin produced goes into alloys.

Lead

Occurrence

The proportion of lead by weight in the Earth's crust is estimated as 0.002%. Galena, lead(II) sulphide, PbS, is the chief ore and is found associated with many other sulphides. Principal sources occur in Australia at Broken Hill (NSW) and Canada which are the two largest exporters, Mexico and the USA. Cerrusite, lead(II) carbonate, $PbCO_3$, and anglesite, lead(II) sulphate, $PbSO_4$, are other ores.

Preliminary treatment of galena

Ores range generally from 3 to 10% in galena content, although some ores are richer and at the other extreme, some at under 1% have been economically worked. The mined ore is crushed and concentrated by froth flotation in which galena rises in the froth formed by air with water containing pine oil and other agents, and is skimmed off. Concentrates may contain 60-80% galena. Before blast-furnace smelting, the ore is spread on a continuous chain grate with limestone and roasted in a sintering machine with the formation of lead(II) oxide and sulphur dioxide:

$$2PbS + 3O_2 \rightarrow 2PbO + 2SO_2$$

Waste gases are trapped as a source of sulphuric acid. A limestone flux is added to obtain the coarse lumps necessary to withstand blast-furnace pressures.

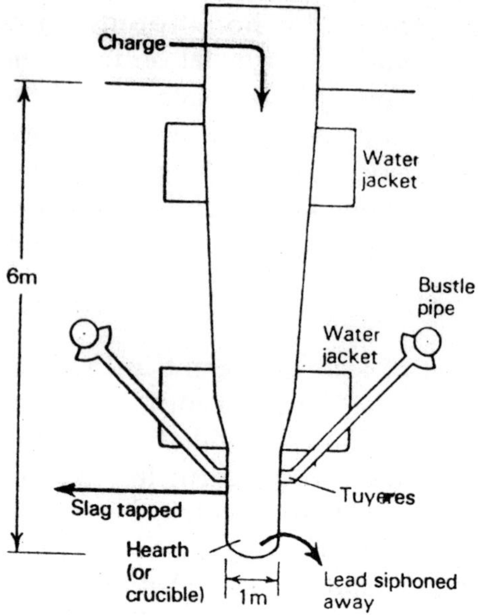

Fig. 3: The blast furnace production of lead (not to scale)

The smelting process

The roasted, sintered concentrate is mixed with coke, limestone to remove siliceous matter as calcium silicate slag, scrap iron or iron oxide ore to give iron for reducing unchanged lead(II) sulphide, and fluxes before heating in a blast furnace, shown in Fig. 3. This may be 3.5-6 m in height, of square section, tapering from about 2 m at the top to 1 m at the base. The temperature reaches 900°C and in the larger furnaces, the shaft is water-cooled. Lead is formed in the hearth -and may be siphoned away. It is called 'base bullion'. From above it, matte (sulphides) and slag (silicates) may be tapped. The reactions are:

$$2C + O_2 \rightleftharpoons 2CO$$

$$PbO + CO \rightarrow Pb + CO_2$$

$$PbS + Fe \rightarrow Pb + FeS$$

$$PbSiO_3 + Fe \rightarrow Pb + FeSiO_3$$

Impurities include copper, arsenic, antimony, tin, zinc, silver, gold and bismuth. These are collected and worked up for marketing.

Refining stage I: copper removal Sulphur is stirred into molten impure lead. Immiscible copper and copper (I) sulphide rise and are skimmed off with the oxide, other sulphides (matte), arsenides and antimonides (speise).

Refining stage II: arsenic, antimony and tin removal, the Harris process Molten lead is pumped through a fused mixture of caustic soda and sodium nitrate or other oxidizing agent for some hours when sodium arsenate, antimonate(V) and stannate(IV) are formed. The elements may be recovered and the alkali regenerated.

Refining stage III: the Parke's process for recovering silver and gold Molten lead is agitated with molten zinc when precious metals are distributed between the almost immiscible metals according to the Partition Law:

$$\frac{\textit{Concentration in zinc}}{\textit{Concentration in lead}} = \textit{constant}$$

(at a given temperature for the same molecular species)

This is called the Partition Coefficient for the system. At 800°C for silver, the Partition Coefficient is 300. The melt is cooled and the zinc content, which solidifies first, is skimmed off. Traces of zinc remaining in the lead are removed by oxidation in air using a reverberatory furnace, by distillation under reduced pressure or by preferential oxidation by chlorine.

The silver-zinc alloy contains up to 10% silver by weight. The precious metals are recovered after lead has been squeezed away in presses. Zinc is distilled off and residual traces of lead are removed by cupellation for about 12 hours in a small reverberatory furnace in which an air blast oxidizes base metals which are tapped as oxides. Silver remains, with any gold and platinum present. Using suitable oxidizing fluxes up to 99.8% silver may be prepared. The hearth is of bone ash, cement or dolomite refractory and the furnaces are gas or oil-fired. Cupellation is followed by electrolytic refining for the silver.

Refining stage IV: removal of bismuth by the Betterton process This is essentially similar in principal to the Parkes' process but a calcium-magnesium alloy is used to recover bismuth. Impurities may be reduced to such a low level that lead is marketed in bulk at 99.9% purity.

Uses

Lead is an extremely dense (11.3 g cm^{-3}), grey metal of low tensile strength, easily worked, being soft, malleable and with a low melting-point (327°C). It is only superficially oxidized at ordinary temperature in air to hydrated oxide (hydroxide) and carbonate. Lead is used for sheathing electrical cables, as sheets for roofing and pipes for plumbing (Latin, *plumbum* = lead). Lead compounds are poisonous.

In hard water districts, lead water pipes soon become coated with lead(II) carbonate and sulphate, preventing uptake of lead compounds in areas served by soft, acidic, moorland water, silicates are added to water to give an insoluble coating of lead(II) silicate. Traces of lead in drinking water have a deleterious effect on health, retarding mental development. For the same reason the use of tetraethyl lead, $Pb(C_2H_5)_4$, as an anti-knock agent in petrol is being phased out in countries of the EEC and elsewhere, particularly in the USA, Lead paints must be avoided where a child might chew the surface, as with cots, toys etc. Lead is highly resistant to sulphuric acid and is used in the Lead Chamber process and as an acid-resisting surface in laboratories. Lead is used to absorb harmful radiation from nuclear reactors and X-ray sources.

Nitrogen, phosphorus, arsenic, antimony and bismuth

Nitrogen

There are many important compounds of nitrogen, finding uses as fertilizers, explosives, dyes and plastics.

Phosphorus

Occurrence

Originally extracted from calcined bone ash, containing about 80% calcium phosphate, by a treatment with sulphuric acid followed by charcoal reduction of the phosphoric acids produced, phosphorus is now produced from natural deposits of calcium phosphate, usually occurring with calcium fluoride in the fluorapatites, $3Ca_3(PO_4)_2 . CaF_2$.

The impure mineral is known simply as phosphate rock. Impurities include carbonates of calcium and magnesium, aluminium oxide and iron(III) oxide. Deposits are worked in the USSR, southern USA, north Africa and on certain Pacific islands. The phosphate content varies widely. Clay and sand are removed by hosing. Fine ore is sintered by heating with fine coke and a little clay and then broken into lumps. About 30-50% of the fluorine escapes as silicon tetrafluoride.

Production

Calcium phosphate (phosphate rock), silica (sand) and carbon (coke) in calculated proportions are heated to about 1500°C in an electric furnace. This process, electrothermal as distinct from electrolytic, has been in use since 1890. Phosphorus, pentoxide, displaced from calcium phosphate by the non-volatile silica, is reduced by carbon to phosphorus vapour, which is led off and condensed under water as the white allotrope:

$$2Ca_3(PO_4)_2 + 6SiO_2 \rightarrow 6CaSiO_3 + P_4O_{10}$$

$$P_4O_{10} + 10C \rightarrow P_4(g) + 10CO(g)$$

tetratomic molecules in the vapour state

The white phosphorus rapidly becomes opaque and brownish-yellow on storage.

The electric power consumed by furnaces varies from 1500 to 30000 kW. Every kg of phosphorus requires 13-15 kW of power to be produced. There are usually three electrodes with a 3-phase AC voltage of 150-300 V and a current of 5000-30000 amperes. One UK phosphorus works is situated alongside an electricity generating station, near to a port. The furnace is a mild steel shell lined with fire bricks, with the hearth and lower parts of the walls protected by carbon blocks.

The three suspended electrodes may be up to 1 m diameter and are maintained at a constant height above the slag, being replaced as they are worn away. The temperature of the melt is maintained by the dissipation of electrical power. The furnace runs until the hearth is worn out, which may be a matter of nine months or several years.

The steel casing is cooled by either a water spray or air. Phosphorus in the emerging gases passes through an electrostatic precipitator and condenses in sprays of warm water. Molten phosphorus, collected under water, is pumped along steam-jacketed pipes for immediate use or to storage vessels, where it is kept under water. The yield is about 90% of the phosphorus available. From the base of the furnace is tapped calcium silicate slag, used for road construction, and the denser ferrophosphorus, FeP or Fe_2P, used in the steel industry to prevent steel-plates from sticking together during pack-annealing. By maintaining the metastable white allotrope of phosphorus at about 270°C for 4-5 days, the stable red modification is produced. The conversion is, of course, exothermic. The element, covered with water, is heated in a steel pot with a firmly secured lid through which passes a safety tube fitted with a short reflux condenser to return escaping phosphorus vapour.

Much excess white phosphorus is distilled off at 400°C and the product ground under water. Boiling with sodium carbonate solution removes the remaining white phosphorus as phosphine, sodium hypophosphite and phosphate. Filtration and vacuum drying yields red phosphorus.

Uses

Much of the phosphorus is converted into orthophosphoric acid and phosphates, for de-rusting and fertilizers. There is much concern about the excessive use of phosphates in farming because of the adverse effect on the natural environment (e.g. algal blooms on lakes) when the excess is leached out of the ground by rain water.

Similarly the use of phosphates in washing powders and detergents has been discontinued in some countries. Phosphorus compounds are used in insecticides, plasticizers, detergents and flotation agents. Phosphorus sesquisulphide, P_4S_3, is produced for 'strike anywhere' watches and red phosphorus for the side of safety matchboxes, the other reactant, potassium chlorate, being placed in the match-head of both types of match. At one time, yellow phosphorus was used in the match industry but constant inhalation of phosphorus vapour by workers produces phosphorus necrosis, 'phossy jaw', and the use of this form of the element has been discontinued.

Arsenic

Occurrence

Thioarsenides are frequently associated with metal ores.

Principal sources of arsenic and its compounds are the by-products of smelting these ores. Arsenopyrite or mispickel, FeAsS, cobaltite, CoAsS, and nickel glance, NiAsS, are examples of thioarsenides. More than enough arsenic trioxide is volatilized during the smelting of copper and lead to satisfy world demand. The sulphides of arsenic have been mined for use as pigments: realgar, As_4S_4, is red, and orpiment, As_4S_6, yellow. Most arsenic is marketed as arsenites and arsenates, formed from sublimed arsenic trioxide. Arsenical compounds are sometimes used as insecticides and for killing potato tops but are giving way to less toxic compounds.

Production

Arsenic trioxide, collected in the course of metallurgical operations, is purified by sublimation in a small reverberatory furnace and reduced by charcoal at red heat in iron or steel retorts:

$$As_4O_6 + 6C \rightarrow \underset{\substack{\text{tetratomic} \\ \text{molecules in the} \\ \text{vapour state}}}{As_4(g)} + 6CO$$

The element distils over and condenses to the crystalline, grey allotrope.

Uses

A little arsenic (0.2%) is used to increase the hardness of lead sheet and is said to cause sphericity in lead shot. Speculum metals, alloys of copper and tin (55: 45 by weight approximately) and arsenic, have a high polish resembling cleaned silver, resist tarnishing and corrosion, and are used as reflectors.

Antimony

Occurrence

The principal ore is antimony trisulphide, stibnite, Sb_2S_3, found in Mexico, Bolivia, South Africa and southern China.

Production

In the preliminary treatment, high-grade ores are heated to about 550°C in a reverberatory furnace when molten stibnite may be run off. Low-grade ore is concentrated by flotation methods. Two methods of antimony smelting may be outlined. In the first, stibnite and scrap iron are heated in a reverberatory furnace, with fluxes to aid the separation of antimony metal and fused iron(II) sulphide matte:

$$Sb_2S_3 + 3Fe \rightarrow 2Sb + 3FeS$$

Otherwise, stibnite is oxidized at 350-450°C with a controlled supply of air and the oxide volatilized. After collection, this is reduced with charcoal or coke for about 10-12 hours using fluxes to hinder the loss of oxide by volatilization:

$$2Sb_2S_3 + 9O_2 \rightarrow Sb_4O_6 + 6SO_2$$

$$Sb_4O_6 + 6C \rightarrow 4Sb + 6CO$$

In both cases, molten antimony is run into moulds. The chief impurities tend to be iron, copper, arsenic and sulphur. The crude metal is heated with the calculated amount of stibnite, liberating more antimony and producing a fused matte of the sulphides of iron and copper, which are skimmed off. Oxidation of arsenic and sulphur is promoted by blowing air through the molten metal.

Uses

Antimony is brittle and is not used by itself. It confers hardness to lead, being used in the production of type metals and bearings.

Bismuth

Occurrence

The principal forms in which bismuth is found are the native element, sulphide ore called bismuth glance or bismuthinite, Bi_2S_3, and associated with these, bismuth ochre, or bismite, Bi_2O_3. The richest sulphide ores occur in Bolivia and the native metal was worked in Saxony. However, most bismuth produced is derived as a by-product from the smelting of lead and copper in Peru, Mexico, Canada and the USA.

Production

Front the sulphide ore, bismuthinite, the extraction closely follows that of antimony, either an oxidation roast in air followed by carbon reduction in a reverberatory furnace or heating with scrap iron and a lime flux:

$$Bi_2S_3 + 3Fe \rightarrow 2Bi + 3FeS$$

$$2Bi_2S_3 + 9O_2 \rightarrow 2Bi_2O_3 + 6SO_2$$

$$Bi_2O_3 + 3C \rightarrow 2Bi + 3CO$$

Various processes are used to recover bismuth from the impurities separated in the smelting of lead and copper. Bismuth of high purity is required for pharmaceutical preparations. Electrolysis is the simplest process in principle with impure bismuth anodes, a bismuth trichloride and hydrochloric acid electrolyte and silver cathodes. Bismuth is deposited at the cathodes from which it is stripped, melted and cast into moulds.

Uses

Like the other elements of this family, bismuth is associated with brittleness and this characteristic limits its use in alloys. Bismuth has a low melting-point (269°C) and it is used to prepare fusible alloys for automatic fire-alarms and safety plugs in boilers.

Oxygen and sulphur

Sulphur

Occurrence

Elemental sulphur is found in the states of Texas and Louisiana in the USA, and in Sicily. It is an important by-product in refineries which process oil of high sulphur content, notably those refining Sahara oil in France. There are many forms of combined sulphur from which sulphuric acid can be made.

The sulphides: galena, PbS, zinc blende, ZnS, and iron pyrites, FeS_2, and the sulphates: gypsum, $CaSO_4 . 2H_2O$, anhydrite, $CaSO_4$, provide sulphur compounds. In Sicily, the element is removed from sulphur-bearing rock by heating. The ore is stacked so as to admit only a controlled supply of air and some of the sulphur ignited. The heat of combustion melts the remainder which is run off. It is later refined by distillation. The USA is the largest producer of sulphur, covering over 90% of the world demand.

Much sulphur is wasted during the combustion of coal in Britain. Fortunately the air of our cities is much cleaner nowadays and we no longer endure the occasional sulphurous smog of the 1950s, and the 'Hollywood film fogs' are no longer a regular feature of London life. However, sulphur dioxide in the waste gases of coal-burning electricity stations is a serious environmental hazard being one of the causes of acid fain.

The provision of electrostatic precipitators etc. is planned but these are very expensive and it may be cheaper to use lower grade imported coal. *Production by the Frasch process* In the USA, sulphur occurs at a depth of about 150 m under soft sand and water. Conventional mining is impossible. In the Frasch process, a hole is bored down to the sulphur bed and lined with a steel casing. The Frasch Sulphur Pump consisting of four concentric steel tubes is lowered into this.

The diameter of the bore-hole is about 300 mm. Super-heated water at about 180°C is forced down the two outer annular spaces and hot compressed air down the axial tube. Sulphur melts under the influence of the superheated water and an emulsion of sulphur, water and air is forced up the remaining annular space. This is shown diagrammatically in Fig. 4. The emulsion is collected in vats where the sulphur solidifying is about 99.8% pure. Alternatively, the element is pumped away molten for immediate use.

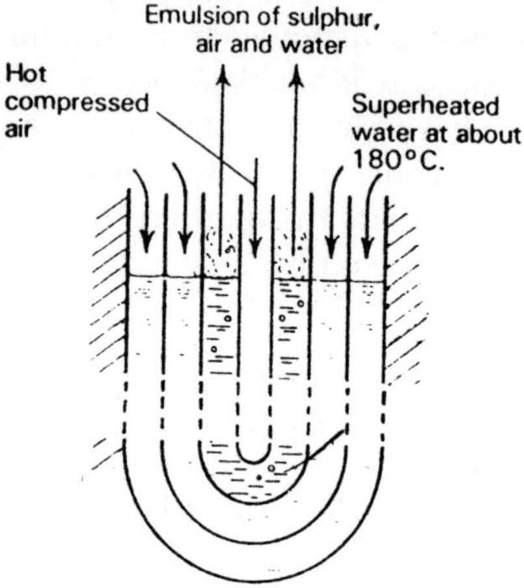

Fig. 4: The Frasch Sulphur Pump (diagrammatic).

Uses

Elemental sulphur is used as a source of sulphur dioxide for the production of sulphuric acid and for making calcium hydrogensulphite for the wood-pulp industry, and directly as the element in vulcanizing rubber and the manufacture of matches and pyrotechnics.

Fluorine, chlorine, bromine and iodine

Fluorine

Fluorine is prepared commercially on a relatively small scale by the electrolysis of potassium fluoride in anhydrous hydrofluoric acid. About 60% by weight of the salt is used. The anodes are of ungraphitized carbon, i.e. without the sheet structure of graphite, because graphite swells and disintegrates under the action of fluorine. The cathode is of steel and the tank, Monel metal. Hydrogen is discharged at the cathode and fluorine at the anode. Volatilized hydrogen fluoride is condensed at -80°C (b.p. +19°C, m.p. -84°C) and the residual 1-2% by absorption in sodium fluoride:

$$HF + NaF \rightarrow NaHF_2$$

Fluorine is used in the concentration of fissionable uranium by the fractional diffusion of uranium hexafluoride and for the production of sulphur hexafluoride, used in the electrical industry, and various fluorides which are used as fluorinating agents.

Bromine

Occurrence

Bromides are distributed widely but in relatively small amounts. There are no important mineral sources, the element being recovered from sea water, inland seas and salt lakes, brines and salt deposits left by the evaporation of seas. The richest inland sea is the Dead Sea, containing 0.56% bromine, from which the element is produced in Israel. Other inland salt waters are found in California, USA, and in the Crimea, USSR. The separation of salt by evaporation of sea water leaves a residual liquor, bitter in taste, called a bittern.

Natural brines containing up to 0.25% bromine occur in Michigan, Ohio and west Virginia, USA. Recovery of bromine from the European saline deposits of Alsace and Stassfurt commences with the recrystallization of salts less soluble than bromides. The mother liquors contain up to 0.5% bromine. Bromine is also recovered from sea water, although it contains only 6.7×10^{-2} g dm^{-3} of the element. Bromine will be extracted in plant located near to the source, where the sea is relatively warm and not diluted by fresh water rivers or sewage.

Production

The extraction of bromine from sea water means processing a large bulk of water: about 20000 kg of sea water are piped for the production of 1 kg of the element. Due to dissolved hydrogencarbonate, carbonates and borates, sea water has a pH = 8.1 at 15°C. Sea water is acidified to pH = 3.5 with sulphuric acid to avoid interaction of halogen in quantity with water and chlorine is introduced. Bromine is liberated by oxidation:

$$2Br^- + Cl_2 \rightarrow Br_2 + 2Cl^-$$

Using the countercurrent principle, air is blown through the liquid and the bromine-laden air is mixed with sulphur dioxide. The resultant hydrobromic and sulphuric acids, together with hydrochloric acid produced in a similar manner from traces of chlorine, are absorbed in water:

$$Br_2 + SO_2 + 2H_2O \rightarrow 2HBr + H_2SO_4$$

Treatment with a countercurrent of chlorine and steam displaces bromine, which distils in the steam and is collected under a saturated solution of bromine in water. Pure bromine is obtained on fractional distillation. Spent water is discharged away from the intake area and the acid generated above used for the initial acidulation of further incoming water. The yield is about 80%. With water containing a relatively high bromide content (more than 0. 1%), the brine is preheated to 90°C and oxidation with chlorine and steam is applied direct without the preliminary concentration effected above. The yield may reach 95% of the bromine content.

As an alternative to sulphur dioxide, aqueous sodium carbonate may be used for absorption, sodium bromide and bromate resulting.

$$3CO_3^{2-} + 3Br_2 \rightarrow 5Br^- + BrO_3^- + 3CO_2$$

Bromine is liberated on acidification with sulphuric acid,

$$5Br^- + BrO_3^- + 6H^+ \rightarrow 3Br_2 + 3H_2O$$

The production of tetraethyl lead from 1,2-dibromoethane (ethylene dibromide) accounted until recently for the larger part (90%) of bromine produced. Tetraethyl lead is an anti-knock agent used to aid combustion in petrol engines but its use is being discontinued because of the harmful effects of traces of lead compounds in the air of towns and cities on health, particularly of the young. There is evidence for the retardation of mental growth. A wide range of inorganic and organic bromine compounds including drugs, dyes and chemical intermediates are used, but the element is only of laboratory importance.

Iodine

Occurrence

Iodine is placed 47th in the order of abundance of elements in the Earth's crust, taking the lanthanides as a single element. Iodine compounds are widely distributed. The principal source of iodine is that dispersed in deposits of sodium nitrate, nitre or caliche, found in northern Chile. It is present chiefly as the mineral lautavite, calcium iodate, $Ca(IO_3)_2$. Caliche nitrate ore stretches in a belt about 500 km long and 25-35 km wide in a desert tableland about 900 m above sea level.

The ore occurs at a depth of 1-15 m and has an iodine content of 0-0.3%. Sea water contains about 0.05 p.p. million of iodine but certain seaweeds (the Laminariaceae) have the power of iodine absorption. Once of major importance, the extraction of iodine from seaweed ash ('kelp') is now of secondary significance. About 75% of iodine originates as caliche, about 18% is produced from brines and less than 7% from seaweed.

Production from caliche

The nitrate-bearing rock is blasted, collected, crushed and

treated with hot water to leach out various soluble salts. The residue is rock, sand and clay. On cooling, sodium nitrate crystallizes out and the mother liquor is recycled for leaching out again. In this way the iodate concentration rises to 6–12 g dm^{-3} iodine. The solution is filtered and an excess of sodium hydrogensulphite solution added.

The liquor is almost neutralized with sodium carbonate solution and an amount of mother liquor, determined by analysis, is added just to complete the generation of iodine. Excess would return some of this into solution as iodide ion, insufficient would leave iodate in solution:

$$3SO_3^{2-} + IO_3^- \rightarrow I^- + 3SO_4^{2-}$$

$$IO_3^- + 5I^- + 6H^+ \rightarrow 3I_2 + 3H_2O$$

when iodine has settled, the supernatant liquor is decanted. Iodine is washed, filtered into canvas bags and pressed free of much of the water. It is sublimed from cement-lined retorts, condensing as crystals in earthenware pipes, each *c.* 0.5 m in diameter and *c.* 1 m long in sets of 6–10. Water condenses and escapes; mineral salts remain behind. The purity is not less than 99%.

Production from iodiferous brine

The iodide ion concentration is 0.003-0.007% by weight, but enough to permit commercial exploitation. The brine is filtered free of oil and mud, treated with sulphuric acid which causes further precipitation, and then sodium nitrite solution is added. The precipitated iodine is absorbed in activated charcoal:

$$2I^- + 2NO_2^- + 4H^+ \rightarrow I_2 + 2NO + 2H_2O$$

Iodine is extracted from the saturated charcoal with caustic soda solution and re-precipitated by the action of acidified sodium dichromate solution:

$$Cr_2O_7^{2-} + 14H^+ + 6I^- \rightarrow 2Cr^{3+} + 3I_2 + 7H_2O$$

It may be purified by melting under sulphuric acid or by steam sublimation.

Uses

Compounds of iodine have very many uses, of which some are the preparation of dye-stuffs, light and heat-sensitive papers, insecticides and vermicides, polaroid film and general organic chemical production. Tincture of iodine (1 g each of potassium iodide, iodine and water in 40 cm^3 of rectified alcohol) is still used as an antiseptic for minor skin cuts and abrasions but it has a strong irritant action. Iodoform is used as an antiseptic.

Chapter 4

Molecular Reactions and Chemical Dynamics

Theoretical explanations of chemical dynamics

Transition-state theory

Transition-state theory is not an exact theory, but is based on a series of approximations. However, it has been useful over a 50-year period. Transition-state theory may also be formulated in terms of changes in thermodynamic properties in the activation process. Since the absorption of light produces excited states of atoms and molecules, photochemistry is really the study of the chemistry of excited states. Electromagnetic radiation in the visible and ultraviolet is generally required to produce chemical reactions because changes in electronic energy levels are required.

Collision theory of bimolecular reactions

Many reactions of atoms or radicals with small molecules in the gas phase have pre-exponential factors in the range $10^{10.5}$ -$10^{11.5}$L mol^{-1} s-1 Since bimolecular reactions of small radicals have zero activation energies, their actual rate constants may be in this range. If the activation energy is zero, we might expect molecules to react on their first collision so that the bimolecular rate constant can be estimated from the collision density Z_{12} between molecules of type 1 and type 2, as calculated for rigid spheres.

Collision density, Z_{12}, given by equation 18.41, is the number of collisions between molecules of type I and type 2 per unit volume per unit time. If we divide by the Avogadro constant N_A, we obtain a reaction rate as it is usually defined

$$\frac{-d[A_1]}{dt} = \frac{Z_{12}}{N_A} = \frac{\pi d_1^{22}(v_{12})}{N_A}\rho_1\rho_2$$

$$= N_A \pi d_{12}^2 [A_1][A_2] \quad (1)$$

where ρ_1 and ρ_2 are number densities, d_{12} is the collision diameter ½$(d_1 + d_2)$, (v_{12}) is the relative speed of molecule A_1, with respect to molecule A_2 and μ is the reduced mass $[(1/m_1) + (1/m_2)]^{-1.}$ Thus, if a bimolecular reaction occurs on the first collision, the rate constant is given by

$$k = N_A\ \pi d_{12}^2 \left(\frac{8k_BT}{\pi\mu}\right) \quad (2)$$

When SI units are used, k has the units of $m^3\ mol^{-1}\ s^{-1}$ and has to be multiplied by $10^3\ L\ m^{-3}$ to obtain the value in the usual units $L\ mol^{-1}\ s^{-1}$. If A_1 and A_2 are identical, a factor of 2 must be included on the right side of equation 2. Equation 2 is only approximate even for bimolecular reactions that occur on the first collision because molecules are not hard spheres that interact only when they touch.

Reactions do not occur unless collisions are sufficiently energetic, both the concept of a collision diameter is a useful one. However, in continuing treatment of bimolecular reactions using collision theory, we will use a quantity S called the reaction cross section, which in area per molecule. To define the reaction cross section we can imagine shooting a beam of A_1, molecules at a volume of A_2 molecules at rest and measuring the number of reactive, collisions that result.

The number of reactive collisions is proportional to the probability P that a given A_2, molecule will be involved in a reactive collision times the cross sectional area, of the beam of A_1 molecules.

Since tire reactive collisions are due to molecules individually of A_1 can also express the number of reactive collisions as the number N_2 of A_2 molecules times the reaction cross section S for the $A_1 + A_2$ reaction. Thus,

$$N_2 S = P\mathcal{A} \text{ or } S = \frac{P\mathcal{A}}{N_2} \quad (3)$$

The reaction cross section depends on the energy of the molecules in the A_2 beam.

More generally the reaction cross section depends on the energy of the collision, since the A_2 molecules are moving too. Rather than assuming that the energy of collision is the sum of the translational energies of spherical molecules A_1 and A_2 at the moment they collide, it is more reasonable to expect that S will be a function of the component of relative kinetic energy along the line of centers at the moment of collision.

This is the energy with which the centers of the two molecules are pushed together.

In simple collision theory it is assumed that this energy must exceed some value ε_0 for a collision to result in reaction. These assumptions lead to the following relations for the reactive cross section as a function of energy.

$$S = 0 \quad \varepsilon < \varepsilon_0 \tag{4}$$

$$S = \pi d^2_{12} \left(1 - \frac{\epsilon_0}{\epsilon}\right) e > e_0 \tag{5}$$

Here d_{12} is the collision diameter for rigid spherical molecules. If the collision energy ε along the line of centers is less than the minimum value ε_0 to cause reaction, the collision cross section is zero; but as the energy increases the reaction cross section S approaches the collision cross section πd^2_{12} for rigid spherical molecules. To derive the equation for the rate constant in terms of the reaction cross section, we return to the experiment in which a beam of A_1, molecules is passed through a layer of A_2 molecules at rest.

The decrease in intensity IA_1, of the A_1 beam due to reactions is given by

$$dI_{A1} = SI_{A1} [A_2] \, dx \tag{6}$$

Here dx is the distance through the layer of A_2 molecules measured in the x direction of the A_1 beam. This equation is analogous to equation 5 used to derive the Beer--Lambert law, except that here we must remember that S is a function of the velocity v of the A_1 molecules. Since $I_{A1} = \nu [A_1]$ and $\nu = d_x/d_f$, equation 6 can be written.

$$-\frac{d[A_1]}{dt} = vS[A_1][A_2] \tag{7}$$

We see that vS is the rate constant for this idealized process this rare, constant has the SI units $m^3 \, s^{-1}$. If gases A_1, and A_2 are reacting in a gas, mixture, in this equation may be interpreted as the magnitude of the relative velocity $v = v_1 - v_2$. However to obtain the rate constant lot a mixture of eases we have to intergate uS over the Maxwell distribution of relative speeds

$$k = \int_{(2\epsilon 0/\mu)1/2}^{\infty} vSf(v)dv \tag{8}$$

Lower limit comes from the fact that $\varepsilon o = \frac{1}{2}\pi v^2$ Since it is more convenient to think of the reaction cross section as a function of energy (equation 4 and 5), equation 8 can be written

$$k = \epsilon \int_{\epsilon 0}^{\infty} \left(\frac{2\epsilon}{\mu}\right)^{1/2} S(\varepsilon)f(\varepsilon)\ d\varepsilon \tag{9}$$

If equation 5 is used for $S(\varepsilon)$, integration leads to

$$k = N_A\ \pi d^2{}_{12} \left(\frac{8k_BT}{\pi\mu}\right) e^{-e0/kBT} \tag{10}$$

Here N_A has been inserted to convert to molar units. The pre-exponential factor in this equation was obtained in equation 2 in a much simpler way. The exponential factor gives the fraction of the collisions in which the energy along the line of centers exceeds ε_0.

Pre-exponential factor in equation 10 is proportional to $T^{½}$ rather than being independent of temperature as in the Arrhenius equation. However, it is not ordinarily possible to distinguish between $AT^{½}e^{-E/RT}$ and $Ae^{-Ea/RT}$ in chemical kinetic data. The exponential term tends to dominate the temperature dependence, and experimental errors in the determination of k limit our ability to determine the functional dependence on T.

According to collision theory, a collision that is sufficiently energetic to supply the activation energy may still fail to produce a reaction if the colliding molecules are not oriented in such a way that they can react with each other, This can be corrected by introducing a steric factor into equation 10. However, this is really not very useful because there is no simple theory for calculating its value. Since there are also very drastic assumptions underlying equation 10, we cannot expect it will yield very accurate results. For example, it ignores the changes in electronic structure that occur in any chemical reaction, and how these changes influence the reactive cross section, To correct this we must look at chemical reactions from the viewpoint of quantum mechanics.

Potential energy surfaces

Kinetic theory is helpful in that it tells us about molecular collisions, but it does not deal with the changes that take place, on a molecular level when reactants, are, converted to products. When two molecules are very close to each other, they cannot be Considered separately because their wave functions overlap. Thus, from the time the reactant molecules are close to each other until the products are well separated, the system is a kind of supermolecule. This supermolecule is different from an ordinary molecule because it is in the process of change, but it is a molecule in the sense that its energy and electron distribution can be calculated for each nuclear configuration by use of quantum mechanics.

According to the Born-Oppenheimer approximation, the electrons move much more rapidly than the nuclei, and so the molecular electronic energy and wave function can be calculated for a given nuclear configuration by use of the electronic Schrobdinger equation. This approximation was used earlier to calculate the electronic potential energy function so that the Schrodinger equation could be used to obtain molecular vibrational energy levels.

If a reaction involves N nuclei, there are $3N$ nuclear coordinates, but the group of nuclei has three translational coordinates of the center of mass and two or three rotational coordinates (about the center of mass) that do not affect the potential energy. Thus, the potential energy is a function of $3N — 5$ nuclear coordinates if the nuclei are constrained to a straight line and $3N— 6$ nuclear coordinates in general. For the simplest type of reaction

$$A + BC \rightarrow AB + C \tag{11}$$

where A, B, and C are atoms, three coordinates are required. It is not possible to plot the potential energy as a function of three coordinates, but if the angle θ of approach of A to BC is fixed, the potential energy of the system can be plotted as a function of R_{AB} and R_{BC} where R is intermolecular distance. Such a plot is shown fit Fig. 1. If R_{AB} is rather large, as on the left face of this diagram, the potential energy is essentially that of the BC molecule. Similarly, the right face gives the potential energy of AB.

Thus, Fig. 1 omits two monotonous valleys to the left and right that extend indefinite distances. Initially, the R_{AB} distance is very large and as A approaches BC, the lowest energy path is given by the dashed line from reactants R to products P. This dashed line gives the minimum energy path, which is sometimes referred to as the reaction coordinate. We will soon see that the configuration of the system does not actually move along the reaction coordinate in the reaction, but the reaction coordinate does help us visualize the surface. The highest point along the reaction coordinate is a saddle point.

At the saddle point, the potential energy is a maximum along the reaction coordinate, but it is a minimum in the direction perpendicular to the reaction coordinate. The reaction system at this point is said to be in the transition state. In Fig. 1, D is a high plateau giving the potential energy of three atoms well separated from each other.

As a first simple example, consider what happens when A approaches a nonvibrating BC molecule along the internuclear axis. The point representing the configuration of the system moves along the minimum energy path, the dashed line in Fig. 1. As R_{AB} decreases, kinetic energy is converted to potential energy as the point representing the system of three nuclei moves up the valley from the left. If there is initially enough kinetic energy for the system to go over the saddle point, AB and C are formed and gain energy as the system goes down the valley to the fight. If the -kinetic energy is too low the system returns down the valley to the left, and we would say that the reactants bounced off each other.

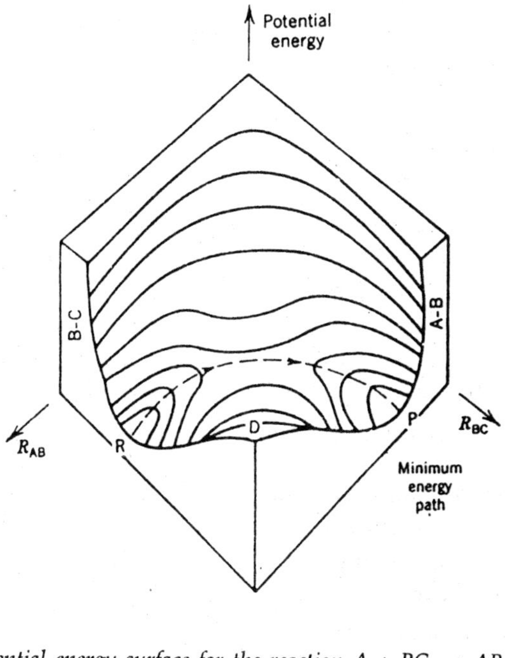

Fig. 1. Potential energy surface for the reaction A + BC → AB + C with the nuclei constrained to a line

Figure 1 applies only when the nuclei are constrained to a line, and the potential energy surface will be different if there is a different angle a of approach. The quantum mechanical calculation of an accurate potential surface for a reaction such as 11 is a difficult process, and so surfaces have been calculated only for a few reactions.

A great deal of attention has been focused on the reaction of a hydrogen atom with a hydrogen molecule:

$$H_A + H_B\,H_C\,H_A \rightarrow H_B + H_C \qquad (12)$$

The potential energy surface for this reaction for θ= 180° is described by means of a contour diagram in Fig. 2. This surface has been calculated using ab initio methods with configuration interaction, and the error at any point on the surface is believed to be less than 0.03 eV (2.9 kJ mol^{-1}). As H_A approaches H_B H_C, along the minimum energy path, the potential energy of the system increases until the saddle point is reached at t.

At this point R_{AB} = R_{BC} = 93 pm, and the potential energy of the system is 0.43 eV (42 U mol^{-1}) the highest along the dashed line. Since the saddle point is 0.43 eV higher than the potential of H_A and HBc at an infinite distance, this energy must be supplied from relative kinetic energy or vibrational energy in order for the reaction to occur. In the upper right corner of Fig. 2, there is a high plateau with energy of 432 kJ mol^{-1}'. This is the energy of three hydrogen atoms infinitely far apart, with respect to separated reactants or products.

Theoretical calculation of a rate constant

Once the potential energy curve has been obtained for various approach angles q, the probability of a reaction for certain initial conditions (relative kinetic energy, vibrational energy and can, in principle, be calculated using the time-dependent Schrodinger equation. This is a

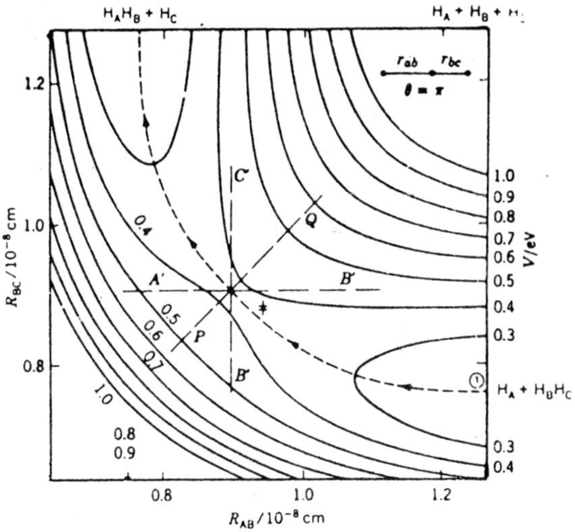

Fig. 2. Potential energy surface for the reaction $H_A + H_BH_C \rightarrow H_AH_B + H_C$ *for a linear approach and departure.*

This is a very difficult calculation, and so classical mechanics is ordinarily used. The force on a particular nucleus is given by the gradient in the potential energy. For example, the component of the force on nucleus i in the x direction is given by

$$F_{x,i} = -\frac{\partial V}{\partial x_i} \tag{13}$$

where V is the potential energy and x_i is the x coordinate of the nucleus. At each instant the system is represented by a point on the surface, and Newton's law $F = ma$ is integrated numerically to obtain the coordinates of the system as a function of time. Calculations can also be made when H_BH_C initially has vibrational motion, and as an approximation this is also treated classically. Rotational energy is not important in these, calculations. These trajectory calculations yield a reaction probability of θ for certain initial condition's and of 1 for other. Figure 3 shows the results of two calculations of this Type for colinear collision. In Fig 3a the H_BH_C molecule is vibrating and H_A, approaches with a certain initial velocity, but reaction does not occur. Note that in this nonreactive, inelastic collision translational energy is converted into vibrational energy in H_BH_C in Fig 3b reaction does occur.

To calculate a rate constant in this way, it is necessary to make a very large number of trajectory calculations with initial states chosen to give a statistically representative sample of possible initial states at the chosen temperature . The initial conditions can be chosen by a Monte Carlo procedure to ensure that the distribution of each initial parameter approaches the correct distribution as the number of calculated trajectories increases. The relative kinetic energies of H_A and H_{BC} are given by the Boltzmann equation. All angles θ of approach have to be included, but most reactive collisions for reaction 12 occur at angles near 180°.

To see how a rate constant can be calculated from a series of trajectory calculations, we need to consider the reaction probability *P*(*b*), often referred to as the opacity function, and the way it is used to calculate a reaction cross section *S*(*v*). The reaction probability *P*(*b*) is simply the fraction of the total number of trajectories at a selected reactant relative velocity and impact parameter b that result in reaction. The reaction probability *P*(*b*) for the H + H_2 reaction for a relative velocity of 1. 17 x 10^6 cm s^{-1} is shown in Fig. 4 as a function of the impact parameter *b*. The reaction probability is greatest for an impact parameter of zero, and it decreases to zero at some finite impact parameter.

The contribution to the reaction cross section *S*(*v*) of collisions with an impact parameter *b* is $2\pi\ bP(b)\ db$, and

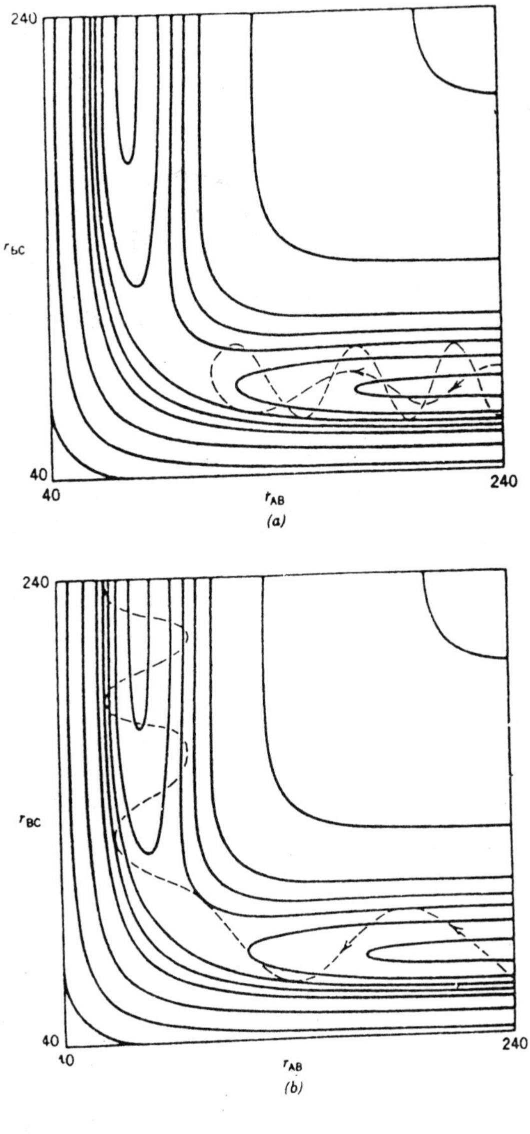

Fig. 3. (a) Trajectory for a nonreactive, inelastic collision of a hydrogen atom and a hydrogen molecule (b) Trajectory for a reactive collision

The cross section can given as

$$S(v) = \int_0^{b_{max}} 2\pi b P(b)\, db \qquad (14)$$

When a chemical reaction occurs in bulk, molecules collide at all possible relative velocities and so the rate constant $k(T)$ at temperature T is made up of a sum of terms for all possible relative velocities, with each weighted by the fraction f1 of collisions with that relative velocity:

$$k(\mathrm{T}) = f_1\, k(v_1) + f_2 k(v_2) + \ldots$$

$$= f1S(v_1)v1 + f_2\, S(v_2)\, v_2 + \ldots \qquad (15)$$

or

$$k(T) = \int_0^{\infty} f(v,\, T)\; vS(v)\; dv$$

where $f(v,T)$ is the Maxwell-Boltzmann distribution for relative velocity v at temperature T. We have already seen a version of this equation in equation 8. In 1965 Karplus, Porter, and Sharma, calculated a very large number of trajectories for the H+ H_2 reaction and found that their results could be expressed by the following Arrhenius equation,

$$k = (4.3 \times 10^{13}\ \mathrm{mole^{-1}\ cm^{3}\ s^{-1}})\ \exp$$

$$\left[-\frac{31000\, J\, mol^{-1}}{(8.314\, J\, K - 1 mol^{-1})T}\right] \qquad (17)$$

This result is in pretty good agreement with experimental results obtained by studying the reaction

$$\mathrm{D + H_2 \rightarrow DH + H} \qquad (18)$$

$$H + para\text{-}H_2 \rightarrow ortho\text{-}H_2 + H \tag{19}$$

and with the value calculated using transition-state theory, which we will discuss in the next section The transition-state rate constant over the same range of temperature is given by

$$k = (7.4 \times 10^{13}\ mol^{-1}\ cm^3\ s^{-1})\ exp$$

$$\left[-\frac{34\,400\, J\, mol^{-1}}{(8.314\, J\, K - 1 mol^{-1})T} \right] \tag{20}$$

Classical calculations have been made on a number of simple reactions, and the general conclusion is that classical calculations can provide an adequate description of the collision dynamics for some purposes. There are quantum mechanical effects in reaction kinetics, and pene:ration into classically forbidden regions ("tunneling") may be important, especially at lower temperatures. (20)

The biggest difficulty in the quantum mechanical calculation of a reaction rate is calculating the potential energy surface with sufficient accuracy. The shape of this surface is extremely important in determining the role of vibrational energy of a reactant in affecting the likelihood of reaction.

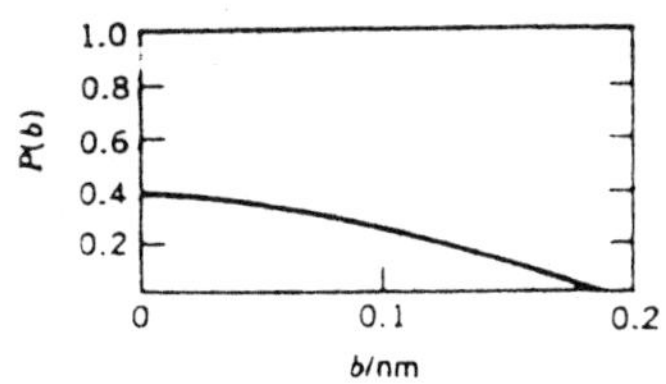

Fig. 4. Reaction probability P(b) for the rection

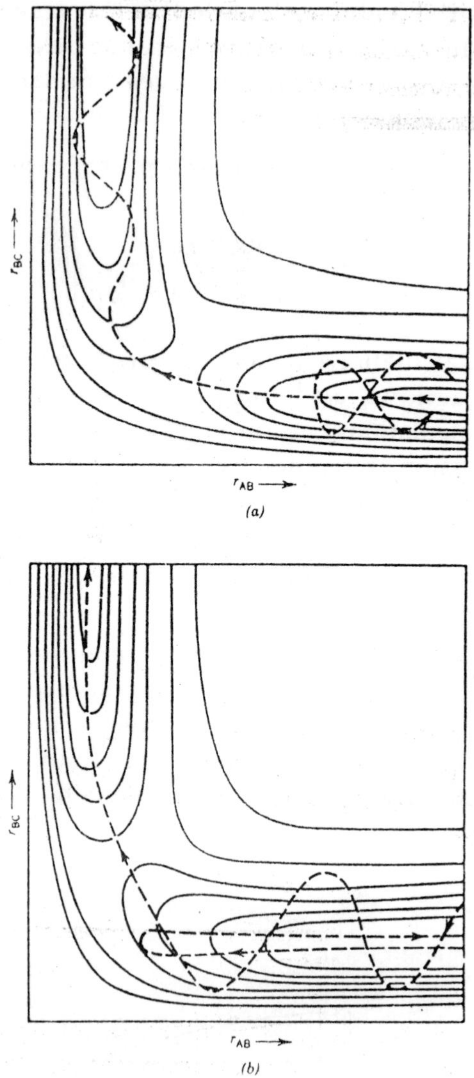

Fig. 5. Attractive potential energy surface for reaction 11(b) Repulsive potential energy surface for reaction 11

Potential energy diagram is symmetrical for the reaction $H_A + H_BH_C = H_AH_B + H_C$, but this is not true in general. The shape of the potential energy surface determines whether translational motion or vibrational motion will be most effective in causing reaction. This behavior can be understood by consideration of the so-called attractive and repulsive potential energy surfaces shown in Fig. 5.

Saddle point occurs early in an attractive surface, as illustrated in Fig. 5a. If BC has a lot of vibrational energy, the reactants may separate without reaction as shown in Fig. 5a. On the other hand, if the system has the same amount of energy, but this energy is mainly relative translational energy, the point representing the system has a higher probability of crossing the barrier. This produces an AB molecule with a significant amount of vibrational energy. Thus, an attractive surface favors reaction when the energy is mainly relative translational motion, and the product molecule is vibrational excited.

When the potential energy surface is of the repulsive type, collisions with relatively high translational energy and little vibrational energy in the original molecule may not be successful. As shown in Fig. 5b, the path of the system is reflected back to products. However, if the original molecule has vibrational energy and the relative kinetic energy is no larger than in the first case, reaction may occur to produce a product with very little vibrational energy. The same surface has been used in discussing both attractive and repulsive situations because a reaction that is attractive in one direction is repulsive in the other.

Transition-state theory

Transition-state theory was developed before much was known about potential energy surfaces, and it, in effect, bypasses the problem of the dynamics of a reactive collision. Nevertheless, it is very useful because quantitative calculations can be made with estimated properties of the transition state for a reaction. The basis of transition-state theory is that it is possible to define a surface in coordinate space and to calculate the flux of trajectories that pass through this surface from the reactant side to the product side without turning back. This flux is identified as the reactive flux. If the potential energy barrier is high, there is no difficulty in locating this surface; it is placed at the top of the energy barrier perpendicular to the reaction coordinate. In addition, transition-state theory is based on the Born-Oppenheimer approximation and the assumption that molecules are distributed among their states according to the Boltzmann distribution.

To derive an expression for the rate constant, we will focus our attention on what happens at the top of the potential energy barrier as shown in Fig. 6, for the reaction A + BC = AB + C. The dashed lines show the quantum mechanical zero-point energies for reactants, products, and the transition state ABC. The transition-state species are those systems within a length 6 along the reaction coordinate, centered at the top of the potential energy barrier.

At chemical equilibrium hall of the ABC complexes are moving forward and hall' are moving backward. It is important to remember that ABC is not an intermediate compound in the reaction, but is a structure that is in the process of falling apart in the direction of the products or in the direction of the reactant, At chemical equilibrium the rate of the reaction in the forward direction can be written either in terms of the macroscopic rate expression for the forward reaction k_f[A][BC] or in terms of the rate that molecules in the transition state are passing over the potential energy barrier to products.

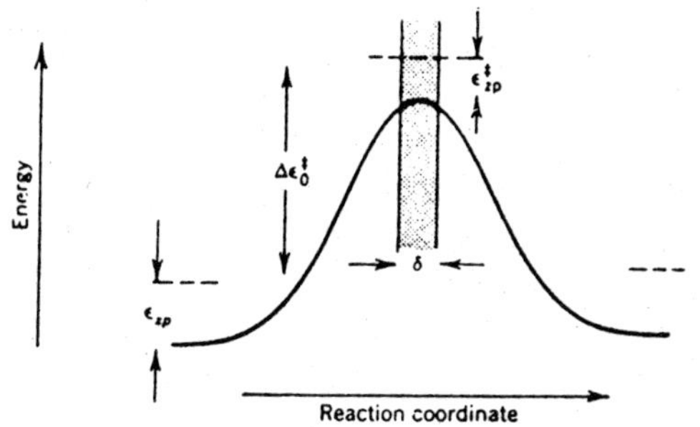

Fig. 6. Potential energy diagram showing the relationship between the height of the barrier in the potential energy surface and the activation energy

$$k_f[A][BC] = \frac{[ABC\pm]\langle\bar{v}_x\rangle}{2\delta} \qquad (21)$$

In this expression is the mean velocity along the reaction coordinate averaged over the 50% of the transition-state species moving in the direction of products. The ratio, the units of reciprocal time.

Similarly, the rate that species in the transition state are passing over the potential energy barrier from product s to reactants can be written in terms of macroscopic kinetics or transition-state theory:

$$k_b[AB][C] = \frac{[ABC\pm]\langle\bar{v}_x\rangle}{2\delta} \qquad (22)$$

It is important to remember that transition-state theory applies to the back reaction as well, and that k_f/k_b = [AB][C]/[A][BC].

We will now concentrate on the calculation of the rate constant for the forward reaction. Using equation 21a, the rate constant for the forward reaction can be calculated from

$$k_f = \left(\frac{v_x}{2\delta}\right)\left(\frac{[ABC^{\ddagger}]}{[A][BC]}\right) \tag{22}$$

Since [ABC]/[A][BC] looks like an equilibrium constant we will represent it by $K^{\ddagger}$ although $ABC^{\ddagger}$ is not a normal molecular species:

$$k_f = \left(\frac{\bar{v}_x}{2\delta}\right) K^{\ddagger} \tag{23}$$

The equilibrium constant $K^{\ddagger}$ for the formation of the transition state can be written in terms of the molecular partition functions q_i for the reactants and q^{ts} for the transition state. Following equation 17.107, $K^{\ddagger}$ can be calculated from properties of the reactant molecules and of the transition state using

$$K^{\ddagger} = \frac{(q^{ts}/V)}{\prod_i q_i/V)} \exp(-\Delta\varepsilon_0^{+}/k_B T) \tag{24}$$

The volume must be included in equation 24 because of the use of concentrations in rate laws.

The last assumption of transition-state theory is that in the transition state, motion along the reaction coordinate is separable. In other words, we will assume that the total partition function q^{ts} for the transition state can be written as the product of the partition function q_x, for the motion along the reaction coordinate and the partition function $q^{\ddagger}$ for all of the other degrees of freedom of the transition state:

$$q^{ts} = q_x q^{\pm} \tag{25}$$

The partition function q_x for one-dimensional motion in the restricted length δ is

$$q_x = \frac{(2\pi\mu_x k_B T)^{1/2}\delta}{h} \tag{26}$$

where, μ_x is the appropriate reduced mass. The average velocity across the top of the barrier in the direction of products is a weighted average for A the velocities of the systems crossing the barrier to products:

Thus, the factor in equation 22 reduces to k_BT/h so that we do not need actual values for δ and μ. The resulting equation for transition-state theory is

$$k = \left(\frac{k_BT}{h}\right)\left[\frac{q^{+}/V}{\prod_i (q_i/V}\right]\exp\left(\frac{-\Delta\in^{+}0}{k_BT}\right) \quad (27)$$

For a bimolecular reaction this yields a rate constant with units of $m^3\ s^{-1}$.

The partition functions for reactants and for the transition state are separable into contributions due to translational motion and to internal motions:

$$q = q_t q_{int} \quad (28)$$

For a unimolecular reaction $q^{\ddagger}_t/V$ cancels $\Pi_i(qi.t/V)$so that the transitionstate expression for the rate constant is (26)

$$k = \left(\frac{k_BT}{h}\right)\left(\frac{q_{int}}{q_{int}}\right)\exp\left(-\frac{\Delta\in_0^{+}}{k_BT}\right) \quad (29)$$

This yields a rate constant in s^{-1}, as it must. This is the rate constant in the high pressure limit, where there are enough collisions to maintain the Boltzmann distribution of A*. However, in the falloff region the rate of formation of vibrationally excited A* molecules is too slow to maintain the Boltzmann distribution. To calculate rate constants in the falloff region it is necessary to calculate the rate constant for the activation process and its reverse. Since these are physical processes they do not depend very strongly on the nature of M.

The classical theory for unimolecular reaction was developed by Rice, Ramsberger, and Kassel. This theory has been improved in several aspects by Marcus, and it is now referred to as the RRKM theory. Since phenomena in the falloff region are quite complicated, we will not be able to discuss them here.

For a bimolecular reaction

$$\frac{q_t^{\ddagger}/V}{\prod_i (q_{i,t}/V)} = \frac{1}{q_t/V} \tag{30}$$

where $q_t/V = (2\pi\mu k_B T)^{3/2}/h^3$ is the partition function per unit volume for the relative motion of the reactants, for which the reduced mass is μ. Thus, the transition- state expression for the rate constant of a bimolecular reaction is

$$k = \left(\frac{kBT}{h}\right)\left[\frac{q_{\text{int}}}{(q_t/V)\prod_i q_{i,\text{int}}}\right]\exp\left(\frac{\Delta \epsilon_0}{k_B T}\right) \tag{31}$$

Internal partition functions for the reactants are calculated in the way described. The calculation of the internal partition function $q^{\ddagger}_{int}$ for the transition state is a problem because in general we do not know the structure of the transition state.

Table 1 Pre-exponential factors for some bimolecular reactions

Reaction	A_{exp} $cm^3\ mol^{-1}s^{-1}$	A_{calc} $cm^3\ mol^{-1}s^{-1}$
$H + H_2 \rightarrow H_2 + H$	5.4×10^{13}	3.5×10^{13}
$Br + H_2 \rightarrow HBr + H$	3×10^{13}	1×10^{13}
$H + CH_4 \rightarrow H_2 + CH_3$	1×10^{13}	2×10^{13}
$H + C_2H_6 \rightarrow H_2 + C_2H_5$	3×10^{12}	1×10^{13}
$CH_3\ H_2 \rightarrow CH_4 + H$	2×10^{12}	1×10^{12}

Pre-exponential factors for a number of bimolecular reactions are given in Table 1.

Pre-exponential factors may be compared with 10^{13} - 10^{15} cm^3 mol^{-1} s^{-1} calculated from simple collision theory.

Last column gives pre-exponential factors calculated from transition state theory using estimated frequencies and geometrical parameters for the activated complex. The activated- complex theory is a considerable advance over simple collision theory.

Thermodynamic formulation of transition-state theory

The quantity in equation 31 has the same general form as the statistical mechanical expression for an equilibrium constant, and this equation can be written as

$$k = \frac{k_B T}{h} K^{\ddagger} \, (c^o)^{1-m} \tag{32}$$

where c° is the standard-state concentration of I mol L^{-1} and in is the order of the reaction, The factor $(c°)^{1-m}$ has to be included because an equilibrium constant is dimensionless, but the units for a rate constant depend on the order of the reaction.

Equilibrium constant *K* may be expressed in terms of an entropy of activation ΔS and an enthalpy of activation ΔH by use of the following familiar relation from thermodynamics:

$$K^{\ddagger} = e^{\Delta S\ddagger/R} e^{-\Delta H\ddagger/RT} \tag{33}$$

Quantities $\Delta S^{\ddagger}$ and $\Delta H^{\ddagger}$ are for standard states (in this case unit concentrations) of the reactants and the activated complex, but we will omit the superscript zeros in order to simplify the notation.

Substituting equation 34 into equation 33 yields

$$k = (c^{o})^{1-m}\, e^{\Delta st/R}\, e^{-\Delta H+/RT} \tag{34}$$

This equation has approximately the form of th Arrhenius equation.

To write equation 35 in the Arrhenius form we need to find the relationship between the Arrhenius activation energy E_a and $\Delta H^{\ddagger}$. The Arrhenius activation energy E_a is defined by

$$E_a = RT^2 \frac{d \ln k}{dT} \tag{35}$$

Since $\Delta U = \Delta H - P\Delta V^{\ddagger}$ at constant pressure

$$E_a = \Delta H+ + RT - P\,\Delta V^{\ddagger} \tag{36}$$

For a reaction with ideal gases, $P\,\Delta V^{\ddagger} = (1 - m)RT$, where m is the order of the reaction. Thus,

$$E_a = \Delta H^{\ddagger} + mRT$$

Substituting this relation into 35 yields

$$k = e^{m}\,(c^{o})^{1-m} \frac{RT}{N_A h}\, e^{\Delta St/R}\, e^{-Ea/RT} \tag{37}$$

So that the pre-exponential factor A in the Arrhenius equation is given by

$$A = e^{m}(c^{o})^{1-m} \frac{RT}{N_A h}\, e^{\Delta St/R} \tag{38}$$

The standard entropy of activation may be calculated from the pre-exponential factor in the Arrhenius equation. The value of $\Delta S^{\ddagger}$ depends on the choice of standard state for bimolecular or trimolecular reactions, but not for unimolecular reactions.

For many unimolecular, bond-breaking gas reactions $\Delta S^{\ddagger} \cong 0$ because the activated complex is much like the original reactants, and there is very little change in configuration in going from the reactants to the activated complex. In this case $e^{\Delta s\ddagger/R} \cong 1$ and, at room temperatures,

$$A = \frac{eRT}{N_A h} = \frac{(2.718)\,(8.31 J\ K\text{-}1\ mol\text{-}1)\,(300\ K)}{(6.02\ 10^{23}\ mol^{-1})\,(6.63\ x\ 10^{-34}\ J\ s)} = 1.70 \times 10^{13}\ s^{-1} \quad (39)$$

for a unimolecular reaction.

If the activation of the molecule involves a rearrangement of atoms or a change in configuration, there will be a change in entropy and is not unity. The values of $\Delta S^{\ddagger}$ are rarely large enough to give a value of more than $10^{2'}$ or less than 10^{-2} to the term and so frequency factors $e(RT/N_A h)e^{\Delta s\ddagger/R}$ for unimolecular reactions may range from about 10^{11} to 10^{15}.

If there is an increase in rotational and vibrational freedom in the activated complex, $\Delta S^{\ddagger}$ is positive, and the pre-exponential factor is larger than 1 .7 x 10^{13} s $^{-1}$. If there is a decrease in rotational and vibrational freedom in the activated complex, $\Delta S^{\ddagger}$is negative, and the pre-exponential factor is smaller than 1.7 x 10^{13} S^{-1}.

Molecular beam experiments

In gas kinetics experiments reactant molecules approach each other on trajectories with random angles and impact parameters and with relative energies that range around those having the highest probabilities at the reaction temperature. Colliding molecules may also have different amounts of vibrational, rotational, and electronic energies. Reaction rate constants arc averages over these various angles of approach, relative energies, and the like.

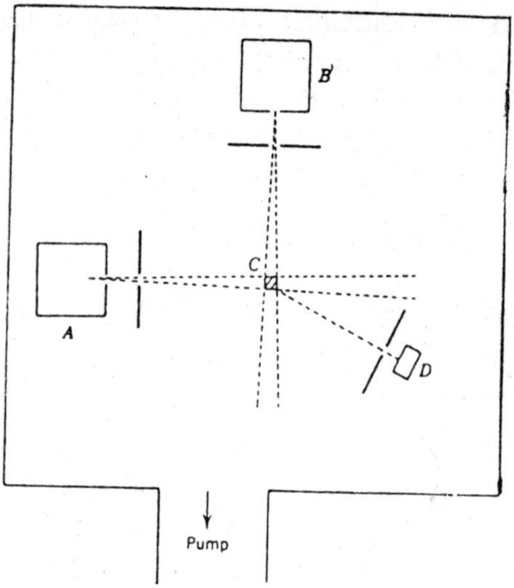

Fig. 7. Schematic diagram for a molecular beam apparatus for studying the reaction of molecules from source B. Products are detected at D.

To get a microscopic view of a bimolecular gas chemical reaction, we would need to carry out the reaction in such a way that reactant molecules are m known quantum states, and the quantum States of product molecules are identified. In the simplest type of apparatus, illustrated schematically in Fig. 7. *A* and *B* are sources of beams of the two reactants that collide in region C. These collisions occur in a chamber evacuated *A* with a high-speed pump so that the only collisions are between the molecules from *A* and *B*. Product Molecules and scattered reactant molecules are detected at *D*.

The effect of changing the angle of approach may be studied by moving *A* or *B*, and the effect of the relative velocity of the reactants may be studied by use of velocity selectors on the beams as they leave *A* and *B*. Sometimes the molecules in molecular beams are put in selected electronic, vibrational, and rotational states so that the effects of these quantum numbers on the observed cross sections can be determined. The term scattering includes three types of phenomena:

1. Elastic scattering in which there are no changes in rotational, vibrational, or electronic quantum numbers in the collision.
2. Inelastic scattering in which there are changes in quantum numbers, but no reaction.
3. Reactive scattering in which new products are formed.

To study the effects of the quantum numbers of molecules in colliding beams the molecules must be prepared in a certain state and there must not be collisions in the beams that would change these quantum numbers.

Molecules can be put into desired vibrational and rotational states by laser excitation. Molecules with dipole moments can be oriented in an in homogeneous electric field so that the effects of molecular orientation in collision can be studied:

$$Na + Cl_2 \rightarrow NaCl + Cl$$

This reaction and some similar reactions have quite large cross sections, which indicates that reaction occurs at distances of the order of 500 pm. The sodium atom loses an electron to the chlorine at a distance of this magnitude so that an ionic intermediate is formed:

$$Na + Cl_2 \rightarrow (Na^{+}.....Cl\text{-}...Cl) \rightarrow Nacl + Cl$$

This is referred to as a "harpooning" mechanism because the electron from the sodium atom can be thought of as a harpoon, and the sodium and chlorine atom can be thought of as being drawn together by electrostatic attraction.

Photochemistry

In studying the electronic spectroscopy of molecules we have seen how the absorption of a photon by a molecule can raise it to a higher energy level, which may have a minimum in the potential energy curve or may not. According to the Franck-Condon principle, the internuclear distance in the molecule is not initially changed by the absorption, and the electric dipole transition moment is proportional to the overlap integral of the initial and final vibrational states. In general selection rules are obeyed, but we must remember that the actual system may be more complicated than the ones for which the selection rules were developed, We will soon see that the selection rule DS = 0, which prohibits transition between states of different multiplicities, is quite important in photoche-mistry. When a molecule absorbs a photon it may lose energy through fluorescence or phosphorescence, but now we will be interested in the fact that chemical reactions may ensue. As pointed out electromagnetic radiation in the visible and ultraviolet is generally required to produce chemical reactions because changes in electronic energy levels are required. More recently it has been found that the absorption of many infrared photons from a high-intensity laser can also cause reaction.

$$A + hv \rightarrow A^* \qquad (40)$$

Thus, a mole of photons can excite a mole of molecules. If the electromagnetic radiation is extremely intense, as in a laser beam, two photons may be absorbed essentially simultaneously:

$$A + 2hv \rightarrow A^{**} \qquad (41)$$

In discussing the intensity of light in connection with the Beer-Lambert law we used to represent the intensity of light of a particular wavelength in terms of energy per unit area per unit time. Because of the two principles in the preceding paragraph we will find it convenient in discussing photochemistry to use Ia to represent the intensity of light absorbed by a system, expressed as the amount per unit volume per unit time, where of course, the amount is expressed in moles of photons. A mole of photons is a convenient unit in photochemistry, and it is frequently referred to as an einstein. The intensity I_a is calculated from the radiant energy of a specific wavelength absorbed per unit volume per unit time by dividing by *NAhv*.

For the photochemical reaction

$$Cl_2 + hv \underset{k_{-1}}{\overset{I_a}{\rightleftharpoons}} 2Cl \tag{42}$$

the rate law is

$$-\frac{d[Cl_2]}{dt} = \frac{1}{2}\ \frac{d[Cl]}{dt} = I_a\ k_{-1}\ [C]^2 \tag{43}$$

Note that each term in this equation has the units mol L^{-1} s^{-1}.If the gas or solution strongly absorbs the light, the reaction will occur only near the surface where the light enters. If the gas or solution is weakly absorbing, reaction will occur throughout the volume, but only a fraction of the light will be absorbed.

The absorption of light can lead to one or more of several processes: fluorescence, phosphorescence, energy transfer to another molecule, or chemical reaction. It is convenient to express the yield of each of these processes as a quantum yield θ. The quantum yield θ is equal to a ratio of rates. For a chemical reaction, it is equal to the reaction rate v divided by the intensity of light absorbed I_a.

$$\phi = \frac{v}{I_a} \tag{44}$$

For fluorescence or phosphorescence it is the rate of emission of radiation divided by the intensity of light absorbed. The quantum yield of chemical reaction is sometimes expressed in terms of a particular reactant or product, but equation 44 has the advantage that the quantum yield applies to a specified reaction.

The quantum yield for fluorescence or phosphorescence is necessarily lessthan unity, and usually much less. The quantum yield for a chemical reaction, however, may be a very large number if the absorption of light produces a radical that starts a chain reaction of a thermodynamically spontaneous reaction the quantum yield for the first step of a step of a chemical reaction, the so-called primary process, is equal to unity or less.

An electronically excited state of a molecule has a different electron distribution and nuclear configuration than the ground state. An electronically excited state of a molecule may be converted spontaneously into more possible products than the ground state because of the additional energy it has.

Before considering chemical reactions we must consider the physical processes that result from the absorption of electromagnetic radiation. The energy absorbed may produce excited molecules that can react chemically, but often the energy is rapidly dissipated as heat.

Quenching

Excited-state molecules may be deactivated rapidly by other molecules in a solution or gas. Such substances are called quenchers, and we will represent them by Q. The quenching of phosphorescence provides a means of determining the rate constant for the reaction of the triplet state molecules and the quencher. The steady-state concentration of triplet molecules in the absence of a quencher. In the presence of a quencher this equation becomes

$$[T1] = \frac{k_{ISC}[S_1]}{k_P + k_{I'SC} + k_Q[Q]} \tag{45}$$

because of the additional pathway for reaction of the triplet-state molecules with the quencher. The quantum yield for phosphorescence in the absence of a quencher and will be represented by p in this section, If the steady-state concentration of T, in the presence of quencher is given by equation 63, the ratio of the quantum yields in the absence and presence of the quencher is given by

$$\frac{\phi \overset{o}{P}}{\phi P} = \frac{k_P + k_{iSC} + k_Q[Q]}{k_P + k_{i'SC}} \tag{46}$$

$$= 1 + k_Q \tau_{T1}[Q]$$

where τ_{T1} the phosphorescence lifetime. The ratio of the intensity $I°p$ of phosphorescence in the absence of quencher to the intensity I_P of phosphorescence in the presence of quencher is proportional to the ratio of the quantum yields and so the preceding equation may be written

$$\frac{I_P^o}{I_P} = I + k_{Q\tau T1}\ [Q] \tag{47}$$

This equation is referred to as the Stern-Volmer equation.

Intermolecular processes

Excited molecules may undergo several different types of reaction. Generally chemical reaction of one excited molecule with another molecule is accompanied by deactivation of the excited molecule, and the excited state is said to be quenched. Quenching an excited molecule (donor *D*) with a second molecule (acceptor A) may result in the electronic excitation of A with concomitant deactivation of *D*. This quenching process is often referred to as electronic energy transfer. Since minute traces of impurities (quenchers) can rapidly deactivate excited molecules, substances and solvents used in photochemical studies must be carefully purified. For example, molecular oxygen reacts rapidly with excited molecules and it is therefore often important to deoxygenate the solutions under study.

Electronic energy transfer process may be classified as radiative transfer, short-range transfer, or long-range resonance transfer. In radiative transfer the donor D emits radiation which is absorbed by the acceptor A:

$$D^* \rightarrow D + hv \tag{48}$$

$$A + hv \rightarrow A^* \tag{49}$$

Short-range energy transfer can occur if the distance between donor and acceptor molecules approaches the collision diameter. A collision is not necessarily required, since the energy transfer can occur at distances slightly greater than the collision diameter. According to the Wigner spin conservation rule, the overall spin angular momentum of the interacting pair of molecules must be unchanged in electronic energy transfer. Like spectroscopic selection rules, this rule is not absolute, but it is an important guide. When only S_0, S_1, and T_i, states are considered, this rule amounts to limiting short-range energy transfer to

$$D^*\ (S_i) + A(S_0) \rightarrow D(S_0) + A^*\ (S_I) \tag{50}$$

$$D^*\ (T_i) + A\ (S_0) \rightarrow D(S_0) + A^*\ (T_I) \tag{51}$$

where D represents the donor molecule and A the acceptor molecule. An example of this second type of short-range energy transfer is

$$\underset{\text{(excited at 366 nm)}}{\text{benzaldehyde}} + \underset{\text{(does not absorb at 366 nm)}}{\text{naphthalene}}$$
$$\rightarrow \text{benzaldehyde} + \underset{\text{(which phosphoresces)}}{\text{naphthalene}} \qquad (52)$$

The efficiencies of these energy transfer processes depend on the relative energies of the states involved. The transfer is relatively efficient if the excited state of D has a greater energy than that of the excited state of A.

Energy transfer may occur in a long-lived collision complex. A molecule M* in an electronically excited state may be a very polarizable species and may, as a result, form a collision complex with a ground state molecule Q. The collision complex M*Q generally has a longer lifetime than the corresponding MQ collision complex. The metastable species M*Q is called an exciplex; if M and Q are the same the resulting excited complex is called an excimer.

The formation of an exciplex can provide a collisional mechanism for the transfer of electronic energy is illustrated in fig 9. When the ground-state molecules M and Q approach, their interactions are repulsive. However, if either M or Q is electronically excited, the energy of the collision pair, M*Q or MQ*, will generally be lower than the separated molecules. At the geometry at which the potential energy surfaces intersect, an internal conversion to the lower surface can occur.

In long-range energy transfer the donor and acceptor molecules are separated by a distance much greater than the collision diameter. The efficiency of the energy transfer depends on the extent of the overlap of the emission spectrum of the donor and the absorption spectrum of the acceptor.

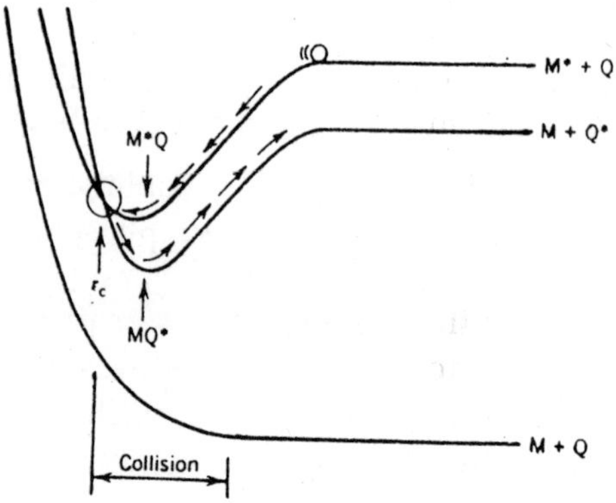

*Fig. 8. Schematic diagram for potential energy surfaces for the formation of exciplexes M*Q and MQ* and internal conversions.*

The energy transfer is not efficient unless the decay process D*D and the excitation process A-A* are allowed electronic transitions. Thus, the most likely long-range energy transfer processes are those given in equations 68 and 69. Under favorable circumstance energy transfer can occur over distances of 5 to 10 nm with rate constants of $10^{10} - 10^{11}$ L mol $^{-1}$ s^{-1}.

The rate constant k_{ET}. 1 for the energy transfer to air acceptor may be obtained by measuring the lifetime of triplet D in the presence and in the absence of A. If the triplet rate of the donor is deactivated solely by phosphorescence, $T_1 \rightarrow S_0$ intersystem crossing, and bimolecular reaction with an acceptor A, the lifetime is given by

$$\tau_{ET} = \frac{1}{k_P + k_{I'SC} + k_{ET}[A]} \tag{53}$$

Since the lifetime T in the absence of the acceptor.

$$\frac{1}{\tau_{ET}} = \frac{1}{\tau} + k_{ET}[A] \tag{54}$$

The fact that second-order rate constants k_{ET} obtained using this relation may equal values expected for diffusion-controlled reactions indicates that energy transfer may take place as rapidly as donor and acceptor can diffuse together.

Chemical reactions and quantum yields

When an excited species undergoes a chemical reaction, we can add another step to the simplified mechanism that we have been discussing. If the excited species (here assumed to be T_1) reacts with reactant R with a bimolecular rate constant k_R it may be possible to evaluate the rate constant

$$T_1 + R \rightarrow \text{stable products} \quad \text{rate} = k_R\,[T_1][R] \tag{55}$$

In the steady state

$$[T_1] = \frac{k_{ISC} I_a}{(k_p + k_{ISC} + k_R[R])(k_{IC} + k_P + k_{ISC})} \tag{56}$$

The quantum yield for the production of stable products is

$$\phi = \frac{k_R[T_1][R]}{I_a} \tag{57}$$

Substituting equation 56 and rearranging yields

$$\frac{1}{\phi} = \frac{k_F + k_{IC} + k_{ISC}}{k_{ISC}}\left(1 + \frac{k_P + k_{ISC}}{k_R[R]}\right) \tag{58}$$

Even for this simplified mechanism there are so many rate constants to be determined before k_R can be obtained that this is not generally practical and so, in further discussions, we will emphasize quantum yields. To determine a quantum yield it is necessary to measure the intensity of the light. This may be done by use of a thermopile, which is a series of thermocouples with one set of junctions blackened to absorb all the radiation, which is then converted into heat. The other set of junctions is protected from radiation.

The amount of radiation may also be measured with a chemical actinometer, in which the amount of chemical change is determined. The yield of the photochemical reaction in the actinometer was determined originally by use of a thermopile.

Reaction 1 has the same value of ϕ from 280 to 300 nm at low pressures and high pressures, in the liquid state or in solution in hexane. The primary process HI + hv = H +1 is followed by the reactions H + HI = H_2 + I and 1 + I_2 thus giving two molecules of HI decomposed for each photon absorbed. Reaction 2, the dimerization of anthracene, has a quantum yield of' unity initially, but the reverse thermal reaction can occur at the product accumulates. In reaction 3 at 366 min the quantum yield is 2 if correction is made for internal screening by the accompanying N_2O_4, which absorbs some light at 366 nm.

The reactions are NO_2 + hv =+ NO_2 + NO_2 = 2NO+ O_2 where the asterisk indicates an excited molecule. At 435 nm and longer wavelengths no reaction occurs when the radiation is absorbed.

Reaction 4 similarly shows a greater quantum yield at shorter wavelengths. This reaction is interesting because at 300°C ϕ has a value of more than 300, indicating that the free radicals that are first produced by the absorption of light are able to propagate a chain reaction at the higher temperatures. At room temperature the reactions involved in the chain do not go fast enough to be detected. The products given in parentheses are present also but in small amounts.

The experimental determination of the quantum yield constitutes an excel tent method for detecting chain reactions. If several molecule of products are formed for each proton of light absorbed, the reaction is obviously a chain reaction in which the products of the reaction are able to promote reaction of other molecules. Reaction 5 is an example of the, fact that the absorption of light in a particular bond does not necessarily cause the rupture of that bond. Acetone like other aliphatic ketones, absorbs ultraviolet light at about 280 nm. The C= O bond which we designate as the chromophore is very strong and does not break to give atomic oxygen. Instead, the absorption energy leads to the cleavage of an adjacent -C- bond that is weaker; thus,

$$\begin{matrix} CH_3 \\ \\ CH_3 \end{matrix} \!\!> C = O + h\nu \rightarrow CH_3 + CH_3C = O \qquad (59)$$

giving a methyl radical and an acetyl radical. The acetyl radical can then decarbonylate, giving CO and $CH_3\cdot$ or it can react with $CH_3\cdot$ to give back acetone. The methyl radicals can couple to form ethane.

In the photolysis of ammonia, reaction 6, hydrogen atoms are split off, and the low yield is probably due to partial recombination of the fragments. The quantum yield varies with pressure and reaches a maximum at 0.6 to 0.7 Pa.

Reaction 7 illustrates a photosensitized reaction. The photodecomposition of oxalic acid, sensitized by uranyl ion, is so reproducible that it is suitable for use as a chemical actinometer. In the uranyl oxalate actinometer the light is absorbed by the colored uranyl ion, and the energy is transferred to the colorless oxalic acid, which then decomposes, The uranyl ion remains unchanged and can be used indefinitely as a sensitizer. The fact that the molar absorption coefficient of uranyl ion is increased by the addition of colorless oxalic acid indicates the formation of a complex. The formation of a chemical complex is often necessary for photosensitization.

Reaction 8 is the best known example of a chain reaction. About 10^5 molecules react for each quantum absorbed. The molecules of hydrogen chloride formed undergo further reaction with the hydrogen and chlorine atoms produced. The measurement of the number of molecules per photon gives a measure of the average number of molecules involved in the chain. Initiation of the reaction with a flash of light can result in an explosively fast reaction.

Reaction 9 illustrates the enhanced reactivity of excited state molecules, The inert complex ion $Co(CN)^{3-}_6$ can be recovered unchanged from boiling water as the potassium salt. However, 313-nm irradiation at 25°C results in aquation to yield $Co(CN)_5(OH_2)^{2-}$ with a quantum yield of 0.30.

Given that the excited state has a lifetime of less than 10^{-10} under the reaction conditions, and that a generous estimate of the rate constant for ground-state aquation would be 10^{-6} s^{-1} the excited state is viewed as being 10^{16} times more reactive than the ground state! Such increases in reactivity are fairly commonly encountered in photochemical work.

The reaction of H_2 and Br_2 to form HBr has been discussed as an example of a chain reaction. At temperatures as low as 200°C the thermal reaction is very slow, but the chain reaction may be initiated photochemically. The empirical rate equation for the photochemical reaction is

$$\frac{d[HBr]}{dt} \quad \frac{k[H_2]I_a^{1/2}}{P^{1/2}\ (1+k'[HBr]/[Br_2])} \tag{60}$$

where I_a is the intensity of absorbed light (in mo L^{-1} s^{-1}) and P is pressure. This steady-state rate equation for be derived by replacing the forward reaction in equation 60 by

$$Br_2 + h\upsilon \rightarrow 2Br \tag{61}$$

with d[Br]/dt = $2I_a$. In the steady state it can be shows that $2I_a$ = 2k –1[Br]^2[M]

so that

$$[Br] = \left(\frac{I_a}{k_{-1}[M]}\right)^{1/2} \tag{62}$$

instead of

$$[Br] = \left(\frac{k_1[Br_2]}{k_{-1}}\right)^{1/2} \tag{63}$$

in the thermal reaction. Using equation 62 in the derivation of the rate equation expected for the photochemical reaction we obtain

$$\frac{d[HBr]}{dt} = \frac{2k_2(l_a/k_{-1})^{1/2}[H_2]}{[M]^{1/2}(1+k_2[HBr]/k_2[Br_2])} \tag{64}$$

which has the same form as equation 60 if [M] is proportional to the total pressure P.

The chain length for a photochemical reaction may be defined in various ways, but for this reaction it is conveniently defined as the number of HBr molecules formed per Br atom produced photochemically. At 500 K and 0.1 bar the chain length is about 100. The quantum yield ϕ is about 200 since each photon absorbed produces two Br atoms. The chain length for the $H_2 + Cl_2$ reaction approaches 10^6 since the chain propagation steps are much faster; as a result a steady state is not reached in this photochemical reaction above 200°C and an explosion occurs.

Ozone layer in the stratosphere

The formation of ozone in the stratosphere is an example of a photochemical stationary state. The ozone formed in this way is important because it absorbs ultraviolet radiation that would otherwise cause damage to life at the surface of the earth. A simplified mechanism for the formation and destruction of ozone in the stratosphere is the following

$$O_2 \xrightarrow[JO_2]{hv} 2O \tag{65}$$

$$O + O_2 + M \xrightarrow{k_2} O_3 + M \tag{66}$$

$$O_3 \xrightarrow[JO_3]{hv} O_2 + O \tag{67}$$

$$O_3 + O \xrightarrow{k_4} 2O_2 \tag{68}$$

where the J's are photodissociation constants defined below. The photolysis reaction in the first step occurs only at wavelengths less than 242 nm. Ozone is formed in the three-body reaction in step 2. The protective role of ozone is due to the third step in which radiation in file 190 to 300-nm range dissociates zone. The fourth step is a slow reaction. The recombination of oxygen atoms with a third body to form O_2 is very slow in the stratosphere and can be neglected. If the absorption of solar radiation in the first and third steps is steady the concentration of ozone rise to a steady level that can be calculated follows The rate of change of ozone concentration is given by

$$\frac{d[O_3]}{dt} = k_2[O][O_2][M] - J_{O3}[O_3] - k_4[O][O_3] \qquad (69)$$

where J_{O3} is the photodissociation coefficient for ozone. The photodissociation coefficient is the probability of dissociation of a molecule per second by light absorption. This coefficient is calculated using

$$J = \int_0^{\infty} \phi_v I_v \sigma_v d_v \qquad (70)$$

where, ϕ_v is the quantum yield of dissociation of the molecule at wavenumber v, Iv is the intensity of sunlight in quanta per unit area per unit time per wavenumber, and σ_v of is the absorption cross section of the molecule at wavenumber v. Thus, I_v has SI units of $m^{-1}s^{-1}$, and J has SI units of s^{-1}.

The rate of change in the concentration of oxygen atoms is assumed to be zero in the steady state since they are present at very low concentrations:

$$\frac{d[O]}{dt} = 2J_{O2}[O_2] - k_2[O][M] + J_{O3}$$

$$[O_3] - k_4[O][O_3] = 0 \qquad (71)$$

Adding, equations 69 and 70 yields

$$\frac{d[O_3]}{dt} = 2J_{O2}[O_2] - 2k_4[O][O_3] \qquad (72)$$

The steady-state concentration of O_3 is given by

$$[O_3]_{ss} = [O_2]\sqrt{k_2 JO_2[M]/k_4 J_{O3}} \quad (73)$$

Since J_{O2}, increases with altitude and $[O_2]$ decreases with altitude, there is a maximum steady-state concentration in the stratosphere, at an altitude of about 20 km. However, equation 73 is only approximate because ozone is involved in more reactions, several of which are catalytic cycles with an environmental impact.

In the late 1960s it was found that nitrogen oxides can catalyze reaction 86:

$$\begin{array}{c} NO + O_3 \rightarrow NO_3 + O_2 \\ NO_2 + O \rightarrow NO + O_2 \\ \hline O_3 + O \rightarrow 2O_2 \end{array} \quad (74)$$

In the mid 1970s it was discovered that chlorine atoms from he Photolysis of chlorofluorocarbons at the level of the ozone layer can also catalyze the decomposition of ozone:

$$\begin{array}{c} Cl + O_3 \rightarrow ClO + O_2 \\ ClO + O \rightarrow Cl + O_2 \\ \hline O_3 + O \rightarrow 2O_2 \end{array} \quad (75)$$

Quantitative calculations based on this mechanism and the discovery of an "ozone hole" over Antarctica have led to international controls on the manufacture of $CFCl_3$ and CF_2Cl_2

Femtosecond transition-state spectroscopy

Development of dye lasers capable of delivering pulses as short as 6×10^{-15} s (6 fs) has made it possible to observe the transition region of a photolysis reaction. In a photolysis reaction the excited molecule ABC* goes through a transition state [A ... BC]* and dissociates into products. Note that [A ... BC]** is different from an unstable intermediate in a reaction because it is in the process of flying apart:

$$ABC^* \rightarrow [A\text{---}BC]'^* \rightarrow A + BC \tag{76}$$

The potential energy curves for A + BC and A + BC* do not have minima, and therefore lead to immediate dissociation. An initial pulse, represented by λ_1 is absorbed at $t = 0$ by the ABC molecule at an internuclear distance close to the potential energy minimum.

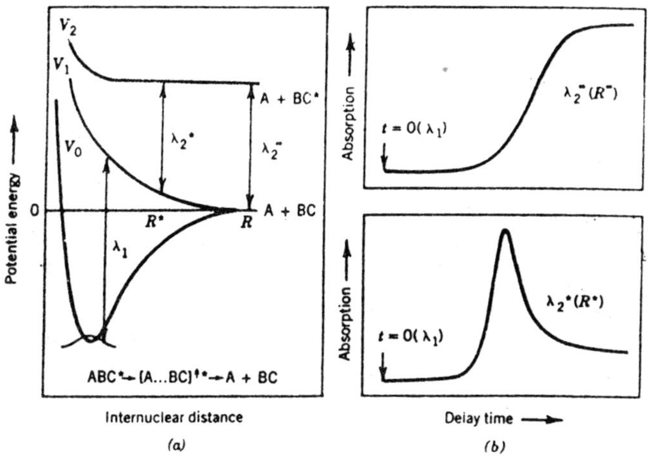

Fig. 9(a). Potential versus internuclear distance R for ABC(V_0), A + BC(V_1) and A + BC. (b) Optical absorption as a function of time in femtoseconds after a pulse at λ_1.

Absorption of a photon by a molecule involves "instantaneous" vertical transition of the molecule from the ground-state potential energy V to an excited-state potential energy V_1. For there to be appreciable absorption, the wavelength of the initial pulse must satisfy the relation $V_1(R_0) - V_0(R_0) \approx hc/\lambda_1$ where R_0 is the internuclear distance at the potential minimum. Classically the ABC molecule dissociate along V_1 with $R \to \infty$ as $t \to \infty$. After a time delay τ the gas is in irradiated with a probe pulse with wavelength λ_2. In the simplest type of experiment, the wavelength λ_2 of the probe pulse is tuned to the absorption wavelength of one of the free fragments. If the probe pulse follows the initial pulse closely, the absorption is negligible initially as shown by the upper plot of optical absorption in Fig. 9b.

The absorption of the probe pulse λ_2 becomes substantial only when the fragments achieve large internuclear distance and are no longer interacting. The build up of the signal from the λ_2^{∞} pulse to half of its maximum takes about 200 fs, as shown by Fig. 9b. Molecules in transition from the initial excited state can be detected by tuning λ_2 away from λ_2. In this second type of experiment, the probe pulse, which is sent at time τ will be absorbed significantly only if the transient-state configuration at time τ is such that where R^* is the internuclear distance at $t = \tau$. The lower absorption plot in Fig. 9b shows that the probe pulse λ^*_2; is absorbed when the transition state has the internuclear distance R^* and that the absorption of λ_2, subsequently falls as the internuclear distance increases and separated products are finally formed, after about 300 fs.

Chapter 5

Oxygen, Oxides and Peroxides

The solubility of oxygen in cold water is only slight, there is enough present to support the life of fish and other aquatic life. Some physical data for oxygen are listed in Table 1.

Table 1

Oxygen to (p block element): physical data

Atomic number	8
Electronic configuration	2, 6
[Core] outer electrons	[He] $(2s)^2$ $(2p)^4$
Relative atomic mass	15. 9994 (16. 00)
Relative molecular mass	31.9988 (32.00)
Atomic radius/nm	0.066
Ionic radius (O^{2-})/nm	0.140
van der Waal's radius/nm	0.140
Electronegativity	3.5
1st ionization energy/kJ mol^{-1}(273K)	1312
2nd ionization energy/kJ mol^{-1} (273 K)	3387
Boiling-point (101 kPa)	-183°C (90K)
Melting-point (101 kPa)	-219°C (54K)
Density/g dm^{-3} (standard conditions)	1.429
Solubility in water/cm^3 dm^{-3}	
(standard condition	48.9
Critical temperature	-119°C(154K)
Critical pressure/(MPa)	5.20

Oxygen occurs freely in the atmosphere of which it comprises 20.8% by volume, roughly one-fifth, and 23.0% by mass, roughly one-quarter. It is the most abundant element in the Earth's crust and with carbon, nitrogen and hydrogen is essential to living matter.

Water contains 89% by mass, or eight-ninths, of combined oxygen and in all it has been estimated hat 49.4% of the Earth's crust consists of the combined element, rocks containing 40–50% by weight. Oxides are extremely important. Ozone, trioxygen, is an allotropic modification of oxygen. It is triatomic, of formula, O_3. It occurs in the higher levels of the atmosphere and reduces the amount of ultraviolet light which penetrates to the surface of the Earth and which would otherwise cause problems of health, notably melanosis, cancer of the skin. He heated nitre and other compounds, including mercury(ii) oxide, collecting the oxygen evolved in bladders. He observed the brilliant combustion of charcoal in what he called 'fire air'. In the meantime, Priestley prepared oxygen on 1st August.

Using a 'burning lens', he focused the rays of the sun on to mercury(ii) oxide (mercurius calcinatus per se), contained in an inverted vessel over quicksilver (mercury). He also produced oxygen by heating minimum (red lead) similarly. He named the gas dephlogisticated air, observed that a candle burned with a vivid flame and recorded that the gas supported the life of a mouse better than did 'common air'. Two months later, in Paris, he related his discovery to Lavoisier who immediately grasped its significance in relation to his own work. Subsequently, combustion, respiration and the calcination of metals were proved by Lavoisier to involve a common process.

He recognized the gas is an element, finally naming it oxygen (Greek, oxys, sour or acid; *genon*, to forth) because products of combustion appeared to him to be acidic. This is only partially true. Some acids were proved to contain no oxygen while some oxides formed alkalis, not acids, with water. Hydrogen cyanide, HCN, was shown by C. L. Berthollet to contain no oxygen. The work of Davy was mentioned in the previous chapter. During this period Lavoisier did much to establish chemistry as a science.

Element No. 8 of the periodic classification

Oxygen is the first member of Group VI of the Periodic Family, which also contains sulphur selenium, tellurium and polonium. There is a greater difference between oxygen and sulphur than between later successive pairs. With atomic number 8, the electronic configuration is 2,6. The most abundant isotope by far has a mass number 16, the nucleus having 8 neutrons as well as 8 protons. For isotopes of mass number 17 and 18, there are 9 and 10 neutrons respectively in addition to 8 protons in the nucleus.

Two atomic mass scales were in use until recently. The physical scale was based on the $^{16}_{8}O$ isotope = 16 exactly and the chemical on the usual isotopic mixture being assigned this value. The interconversion factor is 1.00028.

Physical Atomic mass = Chemical atomic mass $\times$ 1.00028

Elements have slightly larger atomic masses on the Physical Scale. The $^{12}_{6}C$=12 (exactly) scale is now preferred.

Oxygen is second only to fluorine in electronegativity. This leads to its somewhat anomalous behaviour as the first member of Group VI and, in particular, involves it in hydrogen bonding.

Oxygen in combination

Oxygen is in the second period of the Periodic Table, two places from neon. An atom of oxygen may acquire the electronic con figuration of neon in two ways: either by sharing electrons to form covalent or co-ionic bonds, depending on the original distribution of electrons, or by taking two electrons to form an ion.

The simple valency of oxygen in covalent and ionic compounds is two. A marked polarity occurs in covalent compounds d tic to the high electronegativity of oxygen and this is manifested in a dipole moment in unsymmetrical molecules. Binary compounds of oxygen are called oxides.

Oxygen, O_2

A covalent structure could be written in which the octet of electrons corresponding to neon is acquired. Do not think of electrons as localized particles:

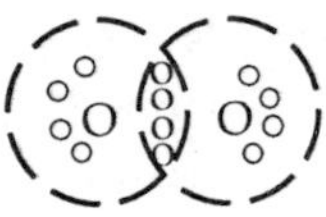

O = O

However, the structure is not straight forward. Oxygen is paramagnetic, as a gas, liquid and solid, and its addition reactions and other characteristics lead to the structure being written:

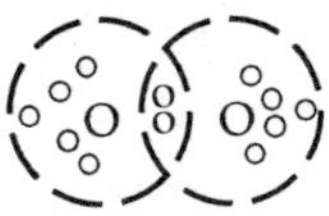

O—O

There are two impaired electrons in the molecule: each atom has an odd electron configuration. This cannot be explained by the basic ideas of the octet theory, but requires the theory of molecular orbitals which may be developed from the atomic orbital theory outlined in the earlier chapters (or fundamentally, developed in parallel to atomic orbital theory).

In this theory the combination of atomic orbitals (or rather the + /—mathematical functions from which they are conceived) results in a re-enforcement of electron density between the approaching atomic nuclei (to form a bonding molecular orbital), or the electron density disappears (an antibonding molecular orbital).

Sigma (σ_s, σ_p) bonds are formed along the internuclear axis (by *s*, *p*, orbitals), while *pi* (π_p) bonds are formed by overlapping of the p_y and p_z orbitals ('sideways on') above and below the internuclear axis but not along it. Bonding orbitals are filled in order of energy, minimum first and bonding orbitals (σ, π) are filled before anti-bonding orbitals (σ^*, π^*), in accordance with Hund's Rule. Compare oxygen, which has two impaired electrons in the πp^* orbitals with fluorine which has no electrons unpaired.

$$O(2s)^2\ (2p)^4 + O(2s)^2\ (2p)^4 \rightarrow O(\sigma_s)^2\ (\sigma_s{}^*)^2\ (\sigma_p)^2\ (\pi_p)^4\ (\pi_p{}^*)^2$$

$$F\ (2_s)^2\ (2_p)^5 + F(2s)^2\ (2_p)^5 \rightarrow F\ (\sigma_s)^2\ (\sigma_s{}^*)^2(\sigma_p)^2\ (\pi_p)^4\ (\pi_p{}^*)^4$$

The net counts of bonding electrons are, respectively, 4(= double bond) and 2(= single bond). If a hypothetical Ne_2 molecule is postulated the next orbital to be filled is the anti-bonding $\sigma_p{}^*$. The resulting net electron count is zero, and there is no bonding, and no molecule. The two unpaired electrons in the $\pi_p{}^*$ orbital are responsible for the paramagnetism mentioned earlier. (The treatment may be extended to comparisons of CO and NO, CN^- and NO^+ etc.)

Oxide type, oxidation state and the periodic table

It was seen how the early attempt to classify certain reactions, which involved the addition of oxygen as oxidation reactions led inexorably to the more sophisticated notion of addition as the loss of electrons by ions, radicals, atoms or molecules (in redox processes) and thence to the use of the term oxidation state (or oxidation number) in developing a systematic nomenclature for inorganic compounds.

The oxidation number of an element is a precise (and limited) term fixed by application of a set of rules. Its use avoids the controversies which might arise over use of the term valency, which attempts to portray by means of theoretical concepts what distinguishes one state of chemical combination of an element from another. This notion will be explained in more detail at appropriate points in the text which follows.

Metals form simple cations by the loss of electrons; those with the stongest tendency to do this are the strongest metals. Hybrid elements, between metals and non-metals in their characteristics, are called metalloids. Consider the oxides in a period of typical elements. Taking Period 3 and writing down the formulae and nature of oxides and the corresponding hydroxides we see that the oxides become progressively less basic and more acidic as the oxidation number increases in steps from I to 7. The oxide bringing out the group oxidation number or valency is chosen in each case:

I	II	III
Na_2O	MgO	Al_2O_3
$NaOH$	$Mg(OH)_2$	$Al(OH)_3$
strong base	strong base	amphoteric

IV	V	VI	VII
SiO_2	P_4O_{10}	SO_3	Cl_2O_7
H_2SiO_3	HPO_3	H_2SO_4	$HClO_4$
weak acid	moderate acid	strong acid	strong acid

In a periodic group or family, the acidity of the corresponding oxide of each element decreases with increasing atomic number:

$$N_2O_5\ P_4O_{10}\ As_4O_{10}\ Sb_2O_5\ Bi_2O_5$$

Bismuth 'pentoxide' which has not been isolated pure is only feebly acidic. With groups of strong metals, the oxides are all basic but the basic nature increases in intensity with increasing atomic number. The change in character of oxides in a group emphasizes the transition from non-metallic to metallic nature, or through lessening degrees of non-metallic character to increasing degrees of metallic nature. When an element forms a series of oxides, the acidity increases with the proportion of oxygen and the valency type becomes more covalent. Manganese affords a good example. The oxidation state of manganese is shown above each formula:

I	II	III
(Mn_2O)	MnO	Mn_2O_3
not ioslated	*basic*	*basic*

IV	V	VI	VII
MnO_2	—	(MnO_3)	Mn_2O_7
amphoteric		*not isolated*	*acidic*

Compounds of MnO_3 are known which indicate that its nature would be acidic. Many other transition metals have similarly graded oxides while others, such as nickel, form only basic oxides. An interesting division of the elements of the Periodic Table, omitting transition and associated elements, may be made using oxide classification. For completeness, various elements not studied in depth here are included but bracketed. Fluorine, which is more electronegative than oxygen, forms oxides (it is thereby reduced), but these are not acidic, basic or amphoteric: therefore, its symbol is ringed. The lines of demarcation are in some cases a matter of opinion.

Arsenic trioxide, AS_4O_6, and arsenious acid, H_3ASO_3, are apparently amphoteric but no cations, AS^{3+}, $As(OH)^{2+}$ or $As(OH)^{2+}$ are known. The formation of $AsCl_3$ with hydrochloric acid is best classed as alcoholic (by analogy to the covalent alcohols which form covalent alkyl halides) and arsenic excluded from the amphoteric region. The next element of the group, antimony, forms one acidic oxide and one amphoteric oxide, while bismuth trioxide is basic and the higher oxide, acidic. The vertical transition from acidic to basic through amphoteric is well shown. The formation of covalent chlorides by the action of hydrochloric acid on oxides and acids is also indicative of the approach of metallic nature. The position of Indium (In) is also arguable.

Reactions in which oxygen is liberated

The affinity of oxygen for elements is generally high and only fluorine will displace it immediately from compounds. The affinity table showing the metal most reactive to oxygen at the head and listing metals in order of decreasing activity is roughly the same as that showing the tendency to produce cations in solution: the electrochemical series.

Metals low in this series (e.g. mercury, silver, platinum and gold) form oxides easily decomposed by heat alone. Mercury(ii) oxide and silver oxide are decomposed to metal and oxygen, mercury volatilizing away:

$$2HgO \rightarrow 2Hg\ (g) + O_2(g)$$

$$2pbO_2 \rightarrow 4Ag + O_2\ (g)$$

Various higher oxides, dioxides and mixed oxides, and peroxides are thermally decomposed to oxygen and a lower oxide. On gentle heating, both red lead and lead dioxide, lead(iv) oxide, break down to yield lead monoxide, lead(ii) oxide, as the solid product:

$$2pb_3O_4 \rightarrow 6pbO + O_2$$

$$2pbO_2 \rightarrow 2pbO + O_2$$

Trimanganese tetroxide is formed from manganese dioxide, manganese(iv) oxide, at red heat:

$$3MnO_2 \rightarrow Mn_3O_4 + O_2$$

The thermal decomposition of salts with anions rich in oxygen. Nitrates, permanganates, dichromates, chlorates, bromates, iodates and peroxodisulphates (persulphates) are decomposed on heating.

At red heat, potassium nitrate yields the nitrite and oxygen:

$$2KNO_3 \rightarrow 2KNO_2 + O_2$$

On warming to 240°C, potassium permanganate readily gives the manganate(vi), manganese(iv) oxide (dioxide) and oxygen:

$$2KMnO_4 \rightarrow K_2MnO_4 + MnO_2 + O_2$$

Potassium dichromate decomposes at white heat to chromate(vi), chromium(iii) oxide and oxygen:

$$4K_2Cr_2O_7 \rightarrow 4K_2CrO_4 + 2Cr_2O_3 + 3O_2$$

Potassium chlorate forms the perchlorate when heated to just above the melting-point, 347°C, and then this decomposes to the chloride and oxygen. The optimum temperature for the first, a disproportionation reaction, is 395°C:

$$4KClO_3 \rightarrow 3KClO_4 + KCl$$

$$KClO_4 \rightarrow KCl + 2O_2$$

The net change is

$$2KClO_3 \rightarrow 2KCl + 3O_2$$

The thermal decomposition of chlorates is hastened by catalysis; various metal oxides may be used, including those of manganese(iv) (manganese dioxide), iron(ii), cobalt(ii), nickel (ii) and copper(ii). The first is well known although the mechanism of the breakdown is not clear. Decomposition is endothermic, starting just above 200°C and being rapid below 300°C, both temperatures being below the melting-point of the chlorate. In this temperature range, manganese(iv) oxide is chemically unchanged. Usually potassium chlorate and granular manganese(iv) oxide are mixed in the proportion of 3:1 by weight, the latter being dried beforehand by heating on an iron tray with a luminous bunsen flame.

Care must be taken: mangnese(iv) oxide containing carbon or antimony sulphide may cause a dangerous deflagration or explosion on heating. A trial portion of the mixture should be heated in an ignition tube before the experiment. The danger of confusing powdered charcoal with manganese(iv) oxide is obvious but accidents have occurred.

The action of hot concentrated sulphuric acid on certain, higher oxides and their anions in salts

Manganese(iv) oxide and chromium(vi) oxide, when separately heated with concentrated sulphuric acid, produce the sulphate of a lower oxidation state, water and oxygen:

$$2MnO_2 + 2H_2SO_4 \rightarrow 2MnSO_4 + 2H_2O + O_2$$

$$4CrO_3 + 6H_2SO_4 \rightarrow 2Cr_2(SO_4)_3 + 6H_2O + 3O_2$$

Potassium dichromate yields oxygen readily on heating with concentrated sulphuric acid,

$$2K_2Cr_2O_7 + 10H_2SO_4 \rightarrow$$

$$4KHSO_4 + 2Cr_2(SO_4)_3 + 8H_2O + 3O_2$$

The decomposition of potassium permanganate must not be attempted: the heptoxide of Mn^{vii} Mn_2O_7, breaks down with explosive violence.

The catalytic decomposition of hypochlorites

Addition of a solution of a cobalt(ii) salt 1, chloride or nitrate, etc.) to a hypochlorite, otherwise chlorate(i), solution results in a brisk evolution of oxygen:

$$2NaClO \rightarrow 2NaCl + O_2$$

It has been suggested that cobalt(ii) oxide or hydroxide formed by the alkali present is converted into the cobalt(iii) compound which decomposes to regenerate the cobalt(ii) compound:

$$CO^{2+} + 2OH^{-} \rightarrow Co(OH)_2$$

$$OCl^{-} + H_2O + 2Co(OH)_2 \rightarrow 2Co(OH)_3 + Cl^{-}$$

$$4Co(OH)_3 \rightarrow 4Co(OH)_2 + O_2 \uparrow + 2H_2O$$

Cobalt(iii) hydroxide has not been isolated.

By the decomposition of peroxides

Hydrogen peroxide readily decomposes on heating the solution to boiling:

$$2H_2O_2 \rightarrow 2H_2O + O_2$$

The decomposition of hydrogen peroxide is influenced by many catalysts; including finely divided metals, gold, silver, platinum especially in the colloidal state, and various oxides. Potassium permanganate in acid solution is used in the titrimetric (volumetric) estimation of hydrogen peroxide, the gaseous product being oxygen,

$$5H_2O_2 + 2KMnO_4 + 4H_2SO_4 \rightarrow$$
$$2KHSO_4 + 2MnSO_4 + 8H_2O + 5O_2$$

Water dripped on to sodium peroxide yields a steady stream of oxygen,

$$2Na_2O_2 + 2H_2O \rightarrow 4NaOH + O_2$$

This can be dangerous because any free metallic sodium present may cause an explosion.

The peroxides of Group II metals decompose on heating. The alternate decomposition and regeneration of barium peroxide was the basis of the now obsolete Brin's process:

$$2BaO_2 \rightarrow 2BaO + O_2$$

By electrolysis

Oxygen appears at the anode during the electrolysis of alkalis, oxoacids and certain of their salts. Pure oxygen is prepared by electrolysis of warm barium hydroxide solution. Any hydrogen which has diffused through the electrolyte into the anode gas may be removed by catalytic oxidation over a red-hot platinum gauze, followed by drying with phosphorus pentoxide.

The reactions of oxygen

Except for the noble (inert) gases and certain very noble metals, such as gold and platinum, oxygen reacts with all elements, although rather special conditions are required, as with the halogens, for some of the reactions. While a reactive element like yellow phosphorus inflames immediately, action is usually slow unless the element is heated, as for copper, or ignited, as for sulphur and hydrogen.

Frequently the reaction is vigorous with the evolution of heat and light. The reaction rate is increased by fine division: heated iron and lead powder inflame, and sulphur vapour from the boiling element burns vigorously.

Covalent oxidation is the addition of a more electronegative element or removal of hydrogen involving covalent (or co-ionic) bonds.

Ionic oxidation is the loss of electrons by atoms, groups of atoms, ions or molecules.

Reduction is the converse of oxidation and takes place at the same time as oxidation.

It is interesting to note that the faster reactions of oxygen are those in which the O—O bond remains unbroken. Examples include the synthesis of Aluminium undergoes surface oxidation in the air, sodium peroxide, Na_2O_2, and barium peroxide, the oxide layer protecting the metal from further BaO_2.

With elements; the synthesis of oxides and peroxides

Hydrogen burns with a pale blue flame to form water,

$$2H_2 + O_2 \rightarrow 2H_2O$$

Groups I and II

While excess of metal in Group I produces the monoxide (e.g. Li_2O),

$$4Li + O_2 \rightarrow 2Li_2O$$

there is an increasing tendency to form a peroxide (e.g. Na_2O_2) or hyperoxides (KO_2, RbO_2, CSO_2). The use of the prefix hyper (Greek = over, excessive) draws attention to the unusually high molar proportion of oxygen. Heated sodium burns in oxygen with a yellow flame to leave yellowish sodium peroxide,

$$2Na + O_2 \rightarrow Na2O_2$$

Potassium burns with a lilac flame to produce the hyperoxide,

$$K + O_2 \rightarrow KO_2$$

For lithium, action stops with the monoxide; sodium goes further with the formation of the peroxide while potassium, rubidium and caesium form the hyperoxide. All are ionic compounds.

Group II metals form the oxide when heated in oxygen. Magnesium and calcium burn with an intense white light,

$$2Mg + O_2 \rightarrow 2MgO$$

$$2Ca + O_2 \rightarrow 2CaO$$

Excess oxygen yields a mixture of oxide and peroxide with heated strontium ($SrO + SrO_2$) and the peroxide with barium (BaO_2).

Beryllium oxide is covalent; the other, ionic

Except for beryllium, metals of' both groups undergo surface oxidation in the cold when exposed to air.

Group III

Finely divided boron burns to the trioxide when heated in air or oxygen,

$$4B + 3O_2 \rightarrow 2B_2O_3$$

Aluminium undergoes surface oxidation in the air, the oxide layer protecting the metal from further action

$$4Al + 3O_2 \rightarrow 2Al_2O_3$$

Groups IV, V and VI

In Period 2, heated charcoal burns with a shower of sparks, but nitrogen will only react reversibly with oxygen at the temperature of an electric arc. Carbon dioxide and nitrogen oxide (nitric oxide) are the respective products:

$$C + O_2 \rightarrow CO_2$$

$$N_2 + O_2 \rightleftharpoons 2NO$$

The yield of nitrogen oxide is only 1.2% by volume at 2000°C and 5.3% at 3000°C.

In Period 3, silicon burns brilliantly when heated in oxygen to form the dioxide. Phosphorus burns, either spontaneously as the white allotrope or when heated as the red, with a brilliant white flame to give trioxide and progressively pentoxide as the proportion of oxygen is increased, and heated sulphur burns with a blue flame to the dioxide,

$$Si + O_2 \rightarrow SiO_2$$

$$4P + 3O_2 \rightarrow P_4O_6$$

$$4P + 5O_2 \rightarrow P_4O_{10}$$

$$S + O_2 \rightarrow SO_2$$

All other elements are oxidized on heating. For example, tin forms the dioxide, tin(iv) oxide, when fused in air or oxygen, and lead, the monoxide, lead(ii) oxide. The valency drop indicates increased metallic character for the metal of higher atomic number, in the same group,

$$Sn + O_2 \rightarrow SnO_2$$

$$2Pb + O_2 \rightarrow 2PbO$$

The halogens (Group VII) require special conditions and as the products are not important to us the reactions are not included.

Transition and associated metals

Steel wool or iron wire, previously heated or dipped into burning sulphur, burns brilliantly with sparks to form tri-iron tetroxide,

$$3Fe + 2O_2 \rightarrow Fe_3O_4$$

On heating in air or oxygen, copper becomes coated with a film of black copper(ii) oxide. The form the pentoxide: metal shows little affinity for oxygen:

$$2Cu + O_2 \rightarrow 2CuO$$

Mercury reacts reversibly according to the temperature and the product, mercury(ii) oxide, is of considerable historical interest in the researches of Scheele, Priestley and Lavoisier. The oxide is decomposed on further heating to higher temperature:

$$2Hg + O_2 \rightleftharpoons 2HgO$$

Combustion of compounds in oxygen

Very many substances burn in oxygen. Consider hydrides. Hydrocarbons form carbon dioxide and water;

$$C_3H_8 + 5O_2 \rightarrow 3CO_2 + 4H_2O$$

Petrol and kerosene, which consist of hydrocarbons, burn similarly. In air, some soot (carbon) may be formed due to incomplete combustion. The exhaust systems of new cars are fitted with catalytic converters to ensure maximum oxidation of the products of petrol combustion and so avoid the awful smog and other harmful effects of excessive use of motor cars in our cities. The USA has led the way with the most stringent measures, now adopted by the EEC, to combat the problem. Leanburn engines are being developed in the UK as a fundamental approach to the problem.

Hydrogen sulphide burns in air and oxygen, with a blue flame, and ammonia in oxygen with a peachcoloured flame:

$$2H_2S + 3O_2 \rightarrow 2H_2O + 2SO_2$$

$$4NH_3 + 3O_2 \rightarrow 6H_2O + 2N_2$$

Addition of oxygen to oxides and hydroxides

Barium oxide forms the peroxide on, heating in air or oxygen:

$$2BaO + O_2 \rightarrow 2BaO_2$$

Oxides of Group I metals, except lithium, also take up oxygen.

Carbon monoxide burns with a blue flame to form the dioxide, nitrogen oxide (nitric oxide) forms the brown dioxide on exposure to air and phosphorus trioxide takes up oxygen or heating to

$$2CO + O_2 \rightarrow 2CO_2$$

$$2NO + O_2 \rightarrow 2NO_2$$

$$P_4O_6 + 2O_2 \rightarrow P_4O_{10}$$

Sulphur dioxide reacts reversibly with oxygen over a heated catalyst of vanadium(v) oxide (pentoxide) or platinized asbestos:

$$2SO_2 + O_2 \rightleftharpoons 2SO_3$$

The hydroxides of certain transition metals in the bivalent state are oxidized in air. Manganese(ii) hydroxide, covalent $Mn(OH)_2$, forms Mn_2O_3 (Mn^{iii}) and MnO_2 (Mn^{iv}), the sesqui-and di-oxides respectively. Iron(ii) hydroxide, $Fe(OH)_2$, forms hydrated iron(iii) oxide, iron(iii) hydroxide $Fe(OH)_3$. Vanadium(ii) and chromium(ii) ions V^{2+} and Cr^{2+}, are very readily oxidized as are iron(ii) ions, Fe^{2+} in neutral solution. The manganese(ii) ion, Mn^{2+}, is the most stable form of combined manganese.

Organic oxidations

Giving one example: ethanal (acetaldehyde) vapour, mixed with air and passed over vanadium(v) oxide (pentoxide) catalyst, is oxidized to ethanoic acid (acetic acid),

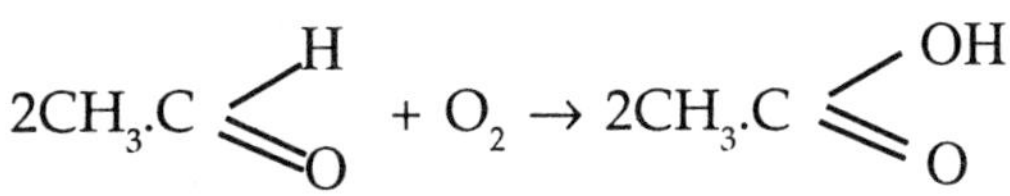

Methods used to prepare oxides

Synthesis

Direct combination of an element with oxygen requires heating or the previous ignition of the element, which may be metallic or non-metallic. Thus, lithium forms the monoxide,

$$4Li + O_2 \rightarrow 2Li_2O$$

The action of steam at red heat

Heated in steam, iron forms tri-iron tetroxide. Carbon monoxide is made industrially (as water gas) by the action of steam on white-hot coke. Hydrogen is also formed:

$$3Fe + 4H_2O \rightleftharpoons Fe_3O_4 + 4H_2$$

$$C + H_2O \rightleftharpoons CO + H_2$$

The thermal decomposition of hydroxides, carbonates and nitrates
These compounds of all but the strongest metals may be decomposed on heating. For calcium:

$$Ca(OH)_2 \rightleftharpoons CaO + H_2O$$

$$CaCO_3 \rightleftharpoons CaO + CO_2$$

$$2Ca(NO_3)_2 \rightarrow 2CaO + 4NO_2 + O_2$$

From other oxides
Higher oxides may decompose on heating, while some lower oxides may take up further oxygen under similar conditions. Lead(iv) oxide (dioxide) and tin(ii) oxide from Group IV provide examples. The former gives litharge, the oxide of bivalent lead, while the latter forms tin(iv) oxide, the oxidation number increasing.

$$2PbO_2 \rightarrow 2PbO + O_2$$

$$2SnO + O_2 \rightarrow 2SnO_2$$

Lead is more metallic in chemical character than tin.

The action of nitric acid
Concentrated nitric acid acts on non-metals and weak metals to form either an oxide, or it' this is soluble, the corresponding acid. Iodine forms iodic acid when boiled with very concentrated nitric acid. Iodine pentoxide is formed on heating:

$$3l_2 + 10HNO_3 \rightarrow 6HIO_3 + 10NO + 2H_2O$$

$$2HlO_3 \rightarrow I_2O_5 + H_2O$$

Photosynthesis

In the light, green plants utilize carbon dioxide, water and light energy to form plant sugars (carbohydrates) with liberation of oxygen. Chlorophyll, the green plant pigment, a complex organic, compound of magnesium is essential to this process which occurs in chloroplasts. Thee are corpuscles containing chlorophyll united to protoplasm. They are usually in the cell layers near the surfaces of leaves. The liberated oxygen has been shown by isotopic ratio studies to come entirely from water and not from the carbon dioxide. Historically, Priestley studied the role of plants in replenishing the air. The work was extended by Ingenhousz, who showed that plants exhale carbon dioxide, and by Jean Senebier.

The uses of oxygen

1. As an aid to respiration, oxygen is used in medical and dental practice. It is supplied where lungs are weak or injury makes artificial respiration impracticable and in the resuscitation of victims of coal gas poisoning and other accidents. It may be used, where appropriate, mixed with anaesthetics. It is essential for high-altitude flying.
2. Oxygen is used in the steel industry for the Bessemer, Open Hearth, Spray and L-D processes in addition to the enrichment of gas used for blast furnaces. An oxygen lance is used to tap a furnace.
3. With combustible gases, ethyne (acetylene), hydrogen, natural gas, it is used for the welding and cutting of ferrous and non-ferrous metals.
4. Chemical processes requiring oxygen include the underground gasification of coal by combustion, the oxidation of hydrocarbons to produce hydrogen / carbon monoxide mixtures for methanol production, the oxidation of ammonia to nitric acid and the oxidation of ethanal (acetaldehyde) to ethanoic acid (acetic acid).
5. Liquid oxygen is used as part of the fuel for the rocket propulsion units required for the various space programmes, and for sustaining life aboard the space stations which orbit the earth.

The allotropy of oxygen: ozone (trioxygen)

Ozone has the formula O_3. Van Marum noticed the distinctive smell of ozone associated with electrical machinery but it was before ozone was recognized as a separate chemical substance by Schonbein. It is called ozone from the Greek, *ozein* = to smell.

Ozone is formed by the action of ultra-violet light on oxygen. It has been estimated that there is a belt of ozone at 25 km height in the atmosphere. The overall quantity of ozone in the atmosphere is estimated as 2 parts per million by volume. There is international concern about the apparent depletion of the ozone layer, after the discovery of a 'hole' in the layer over Antartica. The layer protects life from the harmful effects of excessive ultraviolet radiation, which causes skin cancers.

The loss of ozone is attributed to the action of (among other agents) organic compounds containing fluorine and chlorine (CFCs) used hitherto in aerosols, refrigerators etc., and manufactured because of their apparent chemical inertness. The problem is intensified over the Antarctic subcontinent where the polar vortex reduces mixing of the atmosphere and produces a critical concentration of chlorine atoms. This enables the compounds to penetrate to the stratosphere where subsequent photolytic decomposition produces chlorine atoms, which are regenerated in a chemical cycle that destroys ozone molecules. In the ordinary atmosphere at ground level there is not more than 1.1 parts per million of ozone. Sea air contains no extra ozone whatever the brochures and publicity posters say. While minute traces destroy bacteria, ozone is otherwise poisonous. Ozone melts at -250°C and boils at -112°C. It forms a blue liquid which may explode spontaneously.

With organic matter it explodes. Ozone is the fourth most powerful oxidizing agent, being behind fluorine, oxygen difluoride and atomic oxygen, listed in decreasing power of oxidation.

The formation of ozone from oxygen is endothermic:

$$3O_2 \rightleftharpoons 2O_3; \ \Delta H = +285 \text{ kJ mol}^{-1}$$

It may be shown thermodynamically that ozone is always unstable and breaks down to oxygen. In forming ozone, oxygen exhibits monotropic allotropy or monotropy. The structure is a resonance hybrid:

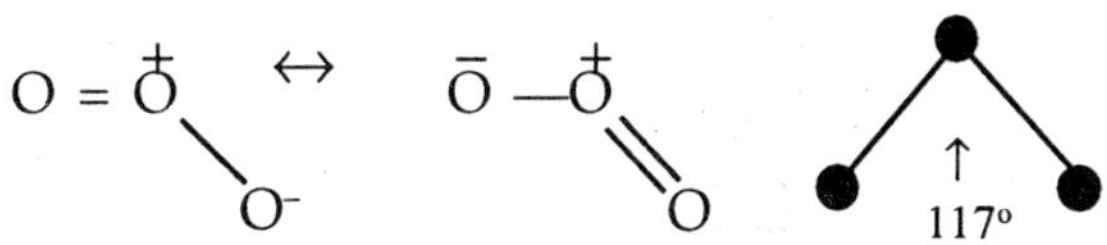

From oxygen using a silent electrical discharge

Oxygen is passed between electrical conductors, but separated from them by the glass walls of the tube acting as an insulator. The conductors are connected to the secondary current of an induction coil and maintained at a high potential difference. Sparking must be avoided because ozone is decomposed rapidly as the temperature is raised. Some ozone is formed,

$$3O_2 \rightleftharpoons 2O_3$$

Two ozonizers are in common use, named after Siemens and Brodie, and shown in Fig. 2. In Siemens' ozonizer, a slow stream of oxygen passes through the annular space between two glass tubes on the outer sides of which are tin foil coatings between which a silent electrical discharge is maintained. In Brodie's ozonizer, the conductor is dilute sulphuric acid which is connected to the potential source by copper wires; the apparatus is necessarily mounted vertically.

The mixture of ozone and oxygen, ozonized oxygen, is led away. But ozone attacks rubber, so that an inert

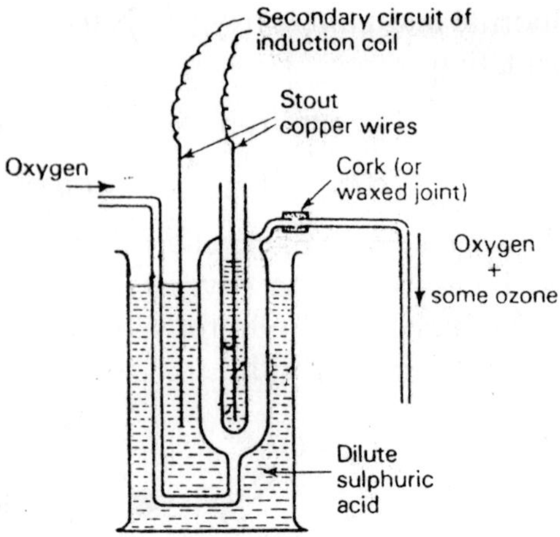

Fig. 1: The preparation of ozonized oxygen in the laboratory.

connection must be made to the delivery tube. These days it is common practice to use ground glass jointed apparatus. The yield is said to be about 10%.

The electrolysis of cold dilute sulphuric acid

Ozonised oxygen is evolved at the anode when icecold dilute sulphuric acid is electrolysed. A high current density is arranged by using a smooth platinum wire electrode, just protruding from the end of a glass tube, into which it is sealed. A 14% yield is said to be obtained.

The reactions of ozone

Ozone is an extremely powerful oxidizing agent. It decomposes to oxygen slowly at room temperature but at an increasing rate as the temperature is raised until at 300°C the decomposition is immediate. Decomposition is catalysed by finely divided platinum, silver(i) oxide and manganese(iv) oxide (dioxide). Except in its reaction with tin(ii) chloride and sulphur dioxide, which are oxidized by oxygen under suitable conditions to form tin(iv) chloride and sulphur trioxide respectively, ozone donates only one oxygen atom per molecule to the substance oxidized:

$$O_3 + X \rightarrow XO + O_2$$

but

$$3SO_2 + O_3 \rightarrow 3SO_3$$

$$3SnCl_2 + 6HCl + O_3 \rightarrow 3SnCl_4 + 3H_2O$$

Iron(ii) chloride is oxidized to iron(iii) chloride in the presence of hydrochloric acid,

$$2Fe^{2+} + 2H^+ + O_3 \rightarrow 2Fe^{3+} + H_2O + O_2$$

and potassium hexacyanoferrate to hexacyano(iii),

$$2Fe(CN)_6^{4-} + H_2O + O_3 \rightarrow 2Fe\ (CN)_6^{3-} + 2OH^- + O_2$$

Dark brown lead(ii) sulphide is oxidized to white lead(ii) sulphate, and hydrogen sulphide to sulphuric acid:

$$PbS + 4O_3 \rightarrow PbSO_4 + 4O_2$$

$$H_2S + 4O_3 \rightarrow H_2SO_4 + 4O_2$$

Potassium iodide in neutral solution liberates iodine, but this reaction is not specific for ozone, being general for oxidizing agents including the halogens, acids and oxides of nitrogen and hydrogen peroxide:

$$2I^- + H_2O + O_3 \rightarrow 2OH^- + O_2 + I_2$$

Except for fluorine, the halogens are displaced from the corresponding hydrides, HCI, HBr, HI,

$$2HBr + O_3 \rightarrow H_2O + Br_2 + O_2$$

Acidified dilute potassium permanganate and dichromate are unchanged by ozone unlike the reactions which occur with hydrogen peroxide.

Mercury loses its mobility when treated with ozone; it 'tails' and smears the glass, the globules and meniscus no longer being visible. This phenomenon is probably due to superficial oxidation affecting the surface tension.

It may be demonstrated by shaking globules of mercury with ozonized oxygen in a corked flask. On adding water, the film of mercury gives way to globules again. Silver foil is attacked by ozone with formation of a brown oxide layer. The tarnishing action of ozone on these two metals distinguishes the gas from pure oxygen.

Of great value to the organic chemist is the oxidation of unsaturated compounds by ozone. Unstable ozonides are formed in anhydrous solvents. Hydrolysis with acid yields aldehydes and ketones. If these are identified, an attempt can be made to picture the original molecule, Consider the normal butenes:

The $>C=C<$ group is replaced by $>C=O \; O=C<$ Oil of turpentine contains two unsaturated hydrocarbons, α-pinene and β-pinene, and may be used to absorb ozone. Cinnamon oil may also be used.

The detection and estimation of ozone

The distinctive odour of ozone is not enough to identify it in small quantities. In any case the sense of small alone can be misleading. The tailing of mercury and the blackening of clean silver together with the liberation of iodine from potassium iodide are characteristic of ozone. However, the sensitive starch-iodide paper test is also positive for other oxidizing agents, halogens, oxides of nitrogen (and nitrous acid) and hydrogen peroxide.

Of these, only ozone and hydrogen peroxide vapour are destroyed by passage along a heated tube. The problem is resolved into the distinction between hydrogen peroxide and ozone in small quantities if the mercury and silver tests are not suitable. Two organic spot reagents are included; they are alcoholic solutions of tetramethyl-parapara-diaminodiphenyl-methane ('tetramethyl base') and benzidine. The detail of other tests is given in practical texts.

Table 3
A comparison of ozone and hydrogen peroxide

	Ozone	*Hydrogen peroxide*
Dilute acidified potassium permanganate	Unchanged	Decolorized
Dilute acidified potassium dichromate and ether	Unchaged	Blue ethereal layer
Tetramethyl base	Violet	Unchaged
Benzidine	Brown	Unchanged
Titranium oxide (dioxide) in concentrated sulphuric acid	Unchanged	Yellow

Quantitative analysis of ozonized oxygen may be in done by absorbing ozone in oil of turpentine. Otherwise, iodine may be liberated from potassium iodide in neutral or somewhat alkaline solution,

$$2KI + O_3 + H_2O \rightarrow 2KOH + O_2 + I_2$$

$$2KOH + I_2 \rightarrow KI + KIO + H_2O$$

Acidification, not done at first because direct oxidation by oxygen must be avoided, liberates iodine which may be titrated against standard sodium thiosulphate solution with starch as indicator,

$$KI + KIO + 2HCl \rightarrow 2KCl + H_2O + I_2$$

Hydrogen peroxide

Hydrogen peroxide was discovered in 1818 by Thenard. It has the formula H_2O_2. Traces are said to be found associated with water and snow, and in the atmosphere. Hydrogen peroxide is used as part of a defence mechanism in plants and animals. Within minutes injured plant cells attack all sorts of chemical intruders (microbes, insects, etc.) by use of oxygen from hydrogen peroxide, liberated by the action of enzymes. This 'nascent' oxygen destroys many organic chemicals, as is seen anyway in the use of hydrogen peroxide in bleaching textiles and wood pulp, and in proprietory lavatory cleaners, marketed as an alternative to hypochlorite solution. In animals white blood cells destroy the protective membranes of invading cells with hydrogen peroxide. Thus it occurs widely in the natural world. When a hydrogen flame impinges on ice, traces of hydrogen peroxide are readily detected by the yellow coloration given with titanium(iv) oxide (dioxide) reagent. It is also formed in traces by the action of ultraviolet light or strong sunlight on water containing dissolved oxygen and directly from the elements in the presence of palladium black as catalyst.

Hydrogen peroxide is prepared in the laboratory by the action of acids on peroxides and commercially by the hydrolysis of peroxodisulphuric acid, produced from fairly con-centrated sulphuric acid by electrolysis, and by auto- oxidation. Hydrogen peroxide is a fairly viscous liquid, 3 density 1.46 g cm^{-3}, with an odour reminiscent of nitric acid, clear and colourless in small quantities but bluish in bulk. It blisters the skin. Hydrogen peroxide is miscible with water and is usually met as a colourless, odourless solution with a sharp metallic taste.

The pure liquid freezes at —1.7°C and explodes when heated to boiling at atmospheric pressure. At a pressure of 9.1 kPa (68 mmHg), hydrogen peroxide boils at 85°C, and at 3.5 kPa (26mmHg), 69°C. It has a higher dielectric constant than water, 89 at 0°C, is a good ionizing solvent and is associated. It dissolves in alcohols and ethers, both oxygen-bearing organic solvents, and in basic liquids, such as aminobenzene (aniline). Hydrogen peroxide is a very weak acid. The thermal decomposition of hydrogen peroxide is exothermic:

$$2H_2O_2 \rightarrow 2H_2O + O_2; \Delta H \quad 2x\ 98\ kJ\ mol^{-1}$$

The pure liquid is fairly stable if stored in a smooth, clean, dark bottle in a cool place. Stabilized 27.5% and 35% solutions should not lose more than 1% of their strength in a year. Storage vessels for concentrated solutions, pipes, etc., should be made from 99.50/6 aluminium. Bleahing vessels for alkaline hydrogen peroxide are made of stainless steel. Polythene and polyvinyl chloride gaskets and flexible hose are used in handling the liquid. Decomposition occurs readily on rough edges, scratch marks and on dust particles. Pure hydrogen peroxide decomposes explosively when heterogeneous catalysts such as finely divided, or colloidal platinum, silver and gold are introduced.

The ions of iron, copper and manganese catalyse decomposition. Rapid decomposition may be brought about by as little as 1 part per million of iron or 0.05 parts per million of copper. Stabilizing agents may be added. These include alcohol, glycerine and N-phenylethanamide (acetanilide). Concentrations greater than 60% will often inflame with organic matter. Solutions of hydrogen peroxide are graded and sold by their 'volume strength'.

The volume strength of a solution of hydrogen peroxide is measured by the number of volumes of oxygen liberated under standard conditions (273 K, 101 kPa) by one volume of solution according to the equation:

$$2H_2O_2 \rightarrow 2H_2O + O_2$$

2 × 34 g hydrogen peroxide yields 22.4 dm^3 of oxygen at STP. If 68 g hydrogen peroxide is made up to 1 dm^3 of aqueous solution, the volume strength is 22.4. Typical strengths are '10 Vol' and '20 Vol'.

10 volume hydrogen peroxide contains

$$\frac{68}{22.4} \times \frac{10}{1} \text{ g dm}^{-3} \text{ hydrogen peroxide, or } 30 \text{ g dm}^{-3}$$

10 volume hydrogen peroxide is a 3% Solution and 20 volume, 60%, by volume, if the density of each solution is assumed to be unity. 'Volumetric strength,' now rarely used, but derived from the analytical method used for the determination of weak solutions: the volume/cm^3 of 0.02 M potassium permanganate solution equivalent to 2 cm^3. Of hydrogen peroxide.

The formula and structure of hydrogen peroxide

The determination of relative molecular mass by the freezing-point method gives the value 34. Two moles of pure hydrogen peroxide on decomposition yield two moles of water and one mole of oxygen. This supports the molecular formula H_2O_2. The decomposition of very nearly pure hydrogen peroxide over mercury by heat and the discovery of the catalytic action of manganese(iv) oxide (dioxide) were achieved by Thenard.

Investigations based on spectroscopy and crystal structure in addition to knowledge of organic peroxide reactions leads to the formulation given overleaf. The H—O bonds are at an angle of 94° to each other and at 97° to the O—O bond which forms the axis of the molecule. The molecule is thus askew with H—O bonds roughly at right angles to each other.

The structure of hydrogen peroxide is probably

```
H—O
   |
   O—H
```

There is some evidence for the existence of another form, but the matter is still in dispute:

Interconversion could occur through ionisation:

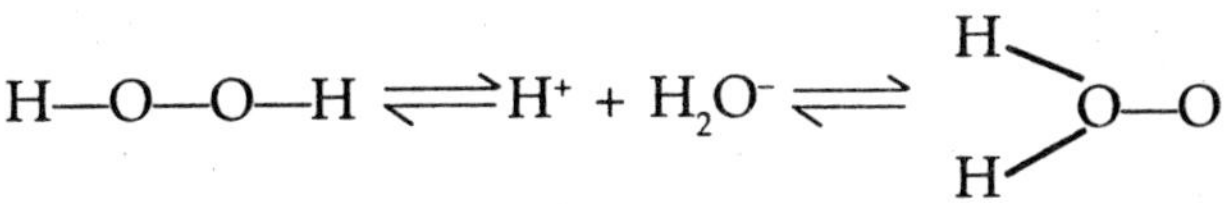

The preparation of hydrogen peroxide

Hydrogen peroxide is formed by the action of dilute acids on Peroxides, usually either barium or sodium peroxide, essentially:

$$2H^+ + O_2^{2-} \rightarrow H_2O_2$$

A dilute solution is obtained.

From barium peroxide

Barium peroxide reacts with sulphuric acid, precipitating barium sulphate. However, commercial barium peroxide is soon coated with insoluble barium sulphate and action ceases. A porous, hydrated form is required.

Commercial barium peroxide is gradually added to a slight excess of ice-cold dilute hydrochloric acid with stirring until it has dissolved. Addition of the equivalent quantity of dilute sodium hydroxide solution, enough to have neutralized the acid taken, results in the precipitation of hydrated barium peroxide, $BaO_2.8H_2O$, which is collected by suction filtration using a Buchner filter funnel:

$$BaO_2 + 2H^+ \rightarrow Ba^2 + H_2O_2$$

$$H_2O_2 + Ba^{2+} + 2OH^- + 6H_2O \rightarrow BaO_2.8H_2O$$

The freshly prepared hydrated barium peroxide is added to ice-cold dilute sulphuric acid. If a slight excess of acid is used, it may be removed by addition of precipitated chalk, or barium carbonate. A solution of hydrogen peroxide remains after filtration:

$$BaO_2.8H_2O + H_2SO_4 \rightarrow BaSO_4 + H_2O_2 + 8H_2O$$

From sodium peroxide

Sodium peroxide is added to a slight excess of icecold dilute sulphuric acid, a little at a time with constant stirring. The decahydrate of sodium sulphate, Na_2SO_4. $10H_2O$, crystallizes out, accounting for some of the water,

$$Na_2O_2 + H_2SO_4 \rightarrow Na_2SO_4 + H_2O_2$$

Alternatively, the calculated quantity of either phosphoric acid or sodium dihydrogenphosphate may be used and partial concentration effected by the crystallization of disodium hydrogenphosphate with twelve molecules of water of crystallization, $Na_2HPO_4.12H_2O$,

$$Na_2O_2 + H_3PO_4 \rightarrow Na_2HPO_4 + H_2O_2$$

$$Na_2O_2 + 2NaH_2PO_4 \rightarrow 2Na_2HPO_4 + H_2O_2$$

The concentration of solutions of hydrogen peroxide

Hydrogen peroxide as ordinarily prepared is readily decomposed on heating, the process being catalysed by rough edges, dust and other surface catalysts. Concentration may be started by careful evaporation in a smooth porcelain or platinum dish over a, water bath to about 60-70°C when effervescence will be observed. Pure hydrogen peroxide has an estimated boiling-point of 144°C at atmosphere pressure, being less volatile than water. A solution of about 45 % concentration is obtained by evaporation in the way described. Unfortunately this method leaves impurities in correspondingly increased concentration, so that the concentrate is more liable to catalytic decomposition. Under reduced pressure, using the apparatus shown in Fig. 2, water is distilled off at 35-40°C under a pressure of about 2 kPa (15 mmHg).

The boiling-point of the contents of the flask increases as concentration proceeds, Distillation is stopped in the range 60-70°C. Further concentration may be achieved by evaporation *in vacuo* over concentrated sulphuric acid in a vacuum desiccator as shown it, Fig. 2. A dish containing the already fairly concentrated solution of hydrogen peroxide rests over concentrated sulphuric acid on a perforated support. The lid and tap are well greased and the desiccator exhausted. Water vapour is absorbed by the sulphuric acid.

Concentrations of more than 91% may be achieved by recrystallization using a freezing mixture of solid carbon dioxide and methanol for cooling.

The commercial production of hydrogen peroxide

Until recently most of the hydrogen peroxide made commercially was electrolytic in origin.

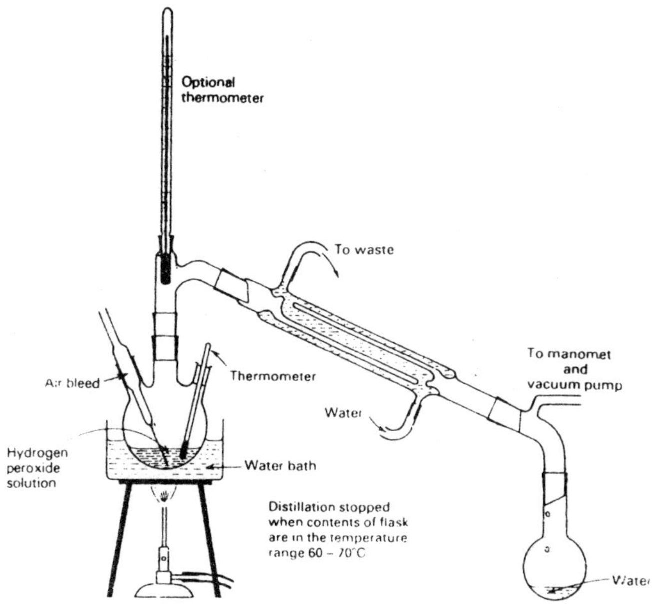

Fig. 2: Concentration of hydrogen peroxide solution under reduced pressure: vacuum distillation of water.

An intermediate, peroxodisulphuric acid, may be prepared by the electrolysis of cold 50% sulphuric acid with a high current density. The anode is bright platinum to keep the oxygen overpotential high and thereby avoid discharge of that gas. The cathode is of graphite or lead. While there is some doubt about the mechanism, peroxodisulphuric acid is formed at the anode. The simplest explanation involves hydrogensulphate ions,

$$HSO_4^- \rightarrow HSO_4 + e\text{-} \quad 2HSO_4 \rightarrow H_2S_2O_8$$

The product is hydrolysed by distillation with sulphuric acid under reduced pressure to give about 30% hydrogen peroxide which may be further concentrated by fractionation:

$$\begin{matrix} O\text{—}SO_2OH \\ | \\ O\text{—}SO_2OH \end{matrix} + \begin{matrix} H_2O \\ \\ H_2O \end{matrix} \rightarrow \begin{matrix} O\text{—}H \\ | \\ O\text{—}H \end{matrix} + 2H_2SO_4$$

In one process, ammonium peroxodisulphate (persulphate), $(NH_4)_2S_2O_8$, is formed by the electrolysis of ammonium sulphate and sulphuric acid and then hydrolysed to ammonium sulphate, sulphuric acid and hydrogen peroxide.

The electrolytic method has now largely been superseded by the 'AO' or auto-oxidation process. The AO process produces hydrogen peroxide from hydrogen and oxygen in the presence of a quinone as catalyst. This yellowish organic solid is converted into a quinol in the course of the reactions. A mixture of solvents is required. The solution is relatively expensive.

The plant is constructed out of 99.5% aluminium and certain aluminium alloys, with stainless steel valves and pumps, etc. The first stage involves the hydrogenation of the quinone to the quinol in the presence of a special catalyst, which is then removed by a highly efficient filtration process,, to avoid later catalytic decomposition of hydrogen peroxide. The second stage involves auto- oxidation of the quinol in the presence of molecular oxygen dissolved in the solution, achieved by passing the solution and air up a packed column. The reactions, using ordinary quinone as the example, because the name of the quinone used has not been disclosed, may e written:

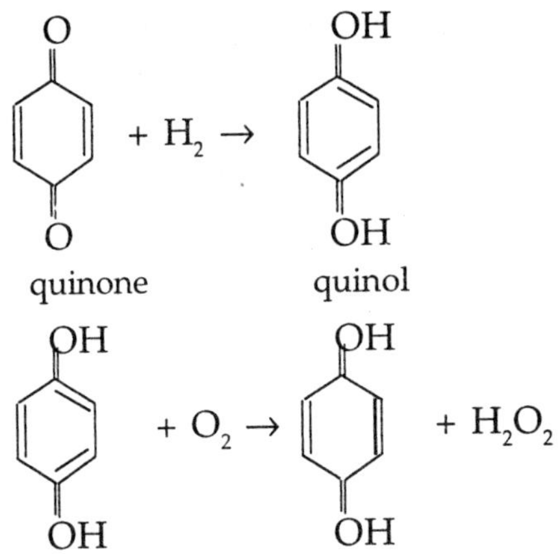

The hydrogen peroxide is extracted by passing demineralized water down the extraction columns in a current counter to the flow of solution. After purification by treatment with various organic solutions, hydrogen peroxide of 18% strength is drawn off. Much of the product is concentrated to high-test product, HTP hydrogen peroxide.

The reactions of hydrogen peroxide solution

The stability of hydrogen peroxide solutions has been discussed already. High-purity hydrogen peroxide has an excellent stability especially if small amounts (less than 0.035%) of organic and/or inorganic stabilizers are added. In addition the pH is adjusted to the range 2-3. The pH of pure aqueous 35% hydrogen peroxide by weight is 5. If the pH is increased, i.e. in alkaline solution, hydrogen peroxide becomes much less stable, decomposition being rapid at pH = 10. Unstabilized hydrogen peroxide, especially in the presence of traces of certain metals, decomposes to water and oxygen:

$$2H_2O_2 \rightarrow 2H_2O + O_2$$

In its reactions hydrogen peroxide may be considered to assume three roles

(a) oxidizing agent

(b) reducing agent, and

(c) acid, forming salts called peroxides.

Hydrogen peroxide as an oxidizing agent

In oxidations, the ion-electron equation of hydrogen peroxide as the oxidant may be expressed:

$$H_2O_2 + 2H^+ + 2e^- \rightleftharpoons 2H_2O$$

Oxidation by hydrogen peroxide is therefore promoted by the presence of acids.

Potassium hexacyanoferrate(ii) (ferrocyanide) solution in the presence' of hydrochloric acid is oxidized to the hexacyanoferrate(iii) (ferricyanide)

$$2Fe(CN)_6^{4-} + H_2O_2 + 2H+ \rightarrow 2Fe(CN)_6^{3-} + 2H_2O$$

Iron(ii) sulphate solution containing dilute sulphuric acid is oxidized to iron(iii) sulphate solution,

$$2Fe^{2+} + H_2O_2 + 2H^+ \rightarrow 2Fe^3+ + 2H_2O$$

Potassium iodide solution, acidified with hydrochloric acid, is oxidized to iodine and potassium chloride,

$$2I^- + H_2O_2 + 2H^+ \rightarrow I_2 + 2H_2O$$

Black lead(ii) sulphide is oxidized to the white lead(ii) sulphate. Both are insoluble. This process may be used in cleaning oil paintings where the white pigment, white lead, $2PbCO_3$. $Pb(OH)_2$, has been darkened by hydrogen sulphide in the air to form lead(ii) sulphide,

$$PbS + 4H_2O_2 \rightarrow PbSO_4 + 4H_2O$$

Sulphur dioxide, bubbled into hydrogen peroxide, is oxidized to sulphuric acid, sulphites are oxidized to sulphates, nitrites to nitrates, and arsenites to arsenates,

$$SO_2 + H_2O_2 \rightarrow H_2SO_4$$

$$SO_3^{2-} + H_2O_2 \rightarrow H_2O + SO_4^{2-}$$

$$NO_2^- + H_2O_2 \rightarrow H_2O + NO_3^-$$

$$AsO_3^{3-} + H_2O_2 \rightarrow H_2O + ASO_4^{3-}$$

Hydrogen peroxide is used in its oxidizing capacity as a mild bleaching agent for wool, hair, teeth, straw, silk, feathers and increasingly, for woodpulp, etc., in the paper industry. It is the standard bleaching agent for both woollen and union blankets. Medicinally, hydrogen peroxide is used as a mild antiseptic.

Hydrogen peroxide as a reducing agent

The ion-electron equation for hydrogen peroxide as a reductant is

$$H_2O_2 + 2OH^- - 2e^- \rightleftharpoons 2H_2O + O_2$$

Alkaline conditions favour this action. In the presence of potassium hydroxide, aqueous potassium hexacyanoferrate(iii) (ferricyanide) is reduced to a solution of the hexacyanoferrate(ii) (ferrocyanide):

$$2Fe(CN)_6^{3-} + 2OH^- + H_2O_2 \rightarrow 2Fe(CN)_6^{4-} + 2H_2O + O_2$$

Iron(iii) salts are reduced when alkali and hydrogen peroxide are added:

$$2Fe^{3+} + H_2O_2 + 2OH \rightarrow 2Fe^{2+} + 2H_2O + O_2$$

Compare this with the reaction under acidic conditions. Potassium permanganate (tetraoxomanganate(vii)), 2 very powerful oxidant, oxides hydrogen peroxide in acid solution: using sulphuric acid, potassium and manganese(ii) sulphates are formed in solution, with oxygen evolution,

$$5H_2O_2 + 2MnO_4^- + 6H^+ \rightarrow 2Mn^{2+} + 8H_2O + 5O_2$$

Manganese(iv) oxide (dioxide) in the presence of sulphuric acid reacts with hydrogen peroxide to give manganese(ii) sulphate and oxygen.

$$MnO_2 + 2H^+ + H_2O_2 \rightarrow Mn^{2+} + 2H_2O + O_2$$

However, under alkaline conditions the reverse happens,

$$Mn^{2+} + H_2O_2 + 2OH \rightarrow MnO_2(s) + 2H_2O$$

Chlorine is reduced to hydrochloric acid, ozone to oxygen and the oxides of silver and gold to the metals by hydrogen peroxide in alkaline solution:

$$H_2O_2 + Cl_2 \rightarrow 2HCl + O_2$$

$$H_2O_2 + O_3 \rightarrow H_2O + 2O_2$$

$$H_2O_2 + Ag_2O \rightarrow 2Ag + H_2O + O_2$$

$$3H_2O_2 + AugO_3 \rightarrow 2Au + 3H_2O + 3O_2$$

Hydrogen peroxide as an acid

Addition of hydrogen peroxide to a saturated solution of barium hydroxide precipitates hydrated barium peroxide, a salt,

$$Ba(OH)_2 + H_2O_2 + 6H_2O \rightarrow BaO_2 . 8H_2O$$

Pure hydrogen peroxide turns litmus solution red but as an acid it is so weak that dilute solutions do not act on indicators,

$$H_2O + H_2O_2 \rightleftharpoons H_3O + HO_2^-$$

Dropwise addition of sodium carbonate solution to hydrogen peroxide solution leads to an evolution of carbon dioxide and the formation of sodium peroxide, Na_2O_2:

$$CO_3^{2-} + H_2O_2 \rightarrow CO_2 + O_2^{2-} + H_2O$$

Addition of hydrogen peroxide to sodium carbonate leads to the evolution of oxygen, because the decomposition of hydrogen, peroxide is catalysed by the alkaline conditions.

The detection and estimation of hydrogen peroxide

Hydrogen peroxide reacts with potassium dichromate (μ-oxobistrioxochromate(vi)) solution acidified with sulphuric acid, and in the presence of ether imparts a blue coloration to the ether. This test is very sensitive and may be used to detect chromates (which on acidification produce dichromates) and dichromates. Covalent chromium peroxide, CrO_5, is formed. It is relatively stable in ether. Otherwise, it is rapidly decomposed:

$$4H_2O_2 + Cr_2O_7^{\ 2-} + 2H^+ \rightarrow 2CrO_5 + 5H_2O$$

$$4CrO_5 + 12H^+ \rightarrow 4Cr^{3+} + 6H_2O + 7O_2$$

A colourless solution of titanium(iv) oxide (dioxide) in fairly concentrated sulphuric acid is converted into the peroxide, TiO_3, by hydrogen peroxide. Even I part of hydrogen peroxide in 180 000 parts of solution gives a light yellow coloration while with increasing concentration the solution deepens in colour and is finally orange-yellow. (This is the most sensitive test.)

Hydrogen peroxide may be estimated volumetrically by direct titration against standard potassium permanganate solution or cerium(iv) sulphate, both in the presence of sulphuric acid:

$$MnO_4^- + 8H^+ + 5e^- \rightleftharpoons Mn^{2+} + 4H_2O \quad Ce4+ +e- \Leftrightarrow Ce^{3+}$$

$$H_2O_2 + 2OH^- - 2e- \rightleftharpoons 2H_2O + O_2$$

(This equation could be written:

$$H_2O_2 - 2^{e-} \rightleftharpoons 2H^+ + O_2)$$

$$\therefore \quad \frac{KMnO_4}{5} \equiv \frac{H_2O_2}{2} \equiv \frac{Ce\ (SO_4)}{1}$$

Hydrogen peroxide is diluted accurately to give reasonable titration readings. Permanganate is run into acidified hydrogen peroxide until a slight permanent pink coloration is obtained:

$$5H_2O_2 + 2MnO_4^- + 6H^+ \rightarrow 2Mn^{2+} + 8H_2O + 5O_2$$

Hydrogen peroxide is run into yellow cerium(iv) sulphate solution. Phenylanthranilic acid is added near the end-point, imparting a mauve coloration, discharged at equivalence:

$$2Ce^{4+} + H_2O_2 \rightarrow 2Ce^{3+} + 2H^+ + O_2$$

In another method, hydrogen peroxide liberates iodine from warm acidified potassium iodide solution,

$$2I^- + H_2O_2 + 2H^+ \rightarrow I_2 + 2H_2O$$

To increase the speed of precipitation, a catalyst of molybdic acid or ammonium molybdate is added, and-iodine titrated immediately against standard sodium thiosulphate (trioxotliiosulphate) solution,

$$I_2 + 2S_2O_3^{2-} \rightarrow 2I^- + S_4O_6^{2-}$$

The concentration/g dm^3 of hydrogen peroxide solution may be converted into the volume strength, which is the volume of oxygen generated under standard conditions by one volume of hydrogen peroxide solution when decomposition occurs according to the equation:

$$2H_2O_2 \rightarrow 2H_2O + O_2$$

Where (2 × 34) g hydrogen peroxide liberates 22.4dm^3 of oxygen tinder STP which is a step from the answer required.

Further, 2 mol hydrogen peroxide liberates 22.4 dm^3

of oxygen under STP, and 1 mol hydrogen peroxide liberates 11.2 dm^3 of oxygen.

Hence volume strength = 11.2 × Molarity.

The uses of hydrogen peroxide

As a mild oxidising agent, hydrogen peroxide has been used to restore oil paintings and other works of art, where lead pigments have been darkened to the sulphide by atmospheric hydrogen sulphide. Lead white pigment (basic lead carbonate $2PbCO_3.Pb(OH)_2$) often used in Italian renaissance drawings, which has gone black may be restored by conversion to the white lead sulphate using ether containing a little hydrogen peroxide, the two immiscible liquids having been well shaken.

Similarly, red lead pigments, which have gone brown or blotchy, are restored, the resulting film of lead being so thin that its whiteness is not seen. Its use as a mild and safe bleach for natural products has been described. Reducing properties make hydrogen peroxide suitable as an antichlor to remove traces of chlorine or hypochlorite (chlorate(i)) remaining after strong bleaching. Hydrogen peroxide is used as an oxidizing agent in the chemical and allied industries. It is especially valuable because the products of decomposition are water and oxygen only.

The oxidation of iron(ii) compounds, phosphorous acid, arsenious acid, and the oxidation of selenium to selenic acid and tellurium to telluric acid are accomplished with hydrogen peroxide. Hydrogen peroxide is used to prepare the peroxides of calcium, strontium, magnesium and zinc on the industrial scale and in the manufacture of peroxo compounds, including sodium peroxoborate and sodium peroxocarbonate, which find widespread application in household detergents as mild and safe bleaching agents.

Hydrogen peroxide is used in large-scale synthesis of organic compounds usually through the intermediary of organic peracids, particularly perethanoic (peracetic) acid, made *in situ* by the action of hydrogen peroxide on an ethanoic acid solution of the compound to be oxidized.

The ease of decomposition of hydrogen peroxide has led to its application in the production of sponge rubber and lightweight concrete where it acts as an aerating agent. Hydrogen peroxide is a useful source of oxygen and has been used with concentrated aqueous sodium permanganate as a propellant for submarine torpedoes and rockets, the mixture generating oxygen and steam. Liquid hydrogen peroxide has been used with liquid hydrazine as a rocket fuel.

Chapter 6

Hydrogen, Hydrides and Water

Molecules of hydrogen are so light that they travel fast enough to escape from the Earth's gravitational field. Spectroscopic studies led to the estimate that 30% by weight of the sun consists of hydrogen. Hydrogen is a colourless, odourless, tasteless, gaseous element. Hydrogen burns in air with an almost invisible blue flame to form water. Physical data are listed in Table 1.

Table 1

Some physical data for hydrogen

Atomic number	1
Electronic configuration	1
Relative atomic mass	1. 00797
Relative molecular mass	2, 01594
Atomic radius/nm	0.029
van der Waals' radius/nm	0.12
Electronegativity	2.1
Ionization energy/kJ mol^{-1}	1312
Boiling point (101 kPa)	-253°C(20K)
Melting point (101 kPa)	-259°C (14K)
Density dm^{-3} (standard conditions)	
Solubility in water/cm^3 dm^{-3}	0.0899
(standard conditions)	21.5
Critical temperature	-241°C(32K)
Critical pressure/Mpa	2.0

About 0.87% by weight of the Earth's crust (the outer 10

miles) consists of hydrogen. Combined hydrogen occurs in all animal and vegetable tissue, and as water. It is estimated that hydrogen accounts for 15.4% of all atoms in the Earth's crust. On this basis it is the second most abundant element, after oxygen. There are more compounds of hydrogen than any other element, including carbon.

The proportion by volume of hydrogen in the atmosphere is less than 1×10^{-6} although it is evolved as 10–30% of natural gas. The element is believed to have the key role in the formation of stars in one theory about the creation of the universe. Although hydrogen had been collected by other workers previously, it was first systematically investi-gated by Henry Cavendish. He observed that the same quantity of hydrogen was evolved when acids (dilute hydrochloric and sulphuric acids) reacted with a fixed weight of a particular metal (zinc, iron, tin) and concluded that the gas came from the metal, calling it 'the inflammable air from metals'. He demonstrated that water was the product obtained on exploding the gas with air or oxygen.

Lavoisier named the gas hydrogen (Greek, *hydorgenon* = water former). Hydrogen was liquefied by Travers. Dewar was believed to have liquefied hydrogen during the previous year but no account of this was published. At ordinary temperatures, sudden expansion from high to low pressure results in a heating effect, but below -80°C the Joule-Thomson cooling effect is obtained. Before commencing liquefaction, hydrogen is cooled in liquid air (-185°C).

Element No. 1 of the periodic classification

Hydrogen is the first element in the Periodic Table. With atomic number = 1, the nucleus of the most abundant isotope is one proton with one valency electron outside it; the mass number is therefore 1.

There is a naturally occurring isotope of mass number = 2. This has a neutron as well as one proton in the nucleus. Of course, the atomic number remains 1 and this isotope occupies the same position (*isos topos*) as ordinary hydrogen in the Periodic Table. However, the doubled mass accentuates the physical difference between the isotopes, which become of great interest and are given special names.

The heavier isotope is called deuterium, $^{2}_{1}H$ = D, and the lighter, sometimes, *protium*, $^{1}_{1}H$ = H. On Earth, the mass ratio H/D is roughly constant (about 6000/ 1) in all sources of hydrogen, so that roughly I in 12 000 atoms of hydrogen is deuterium. Hydrogen, therefore, consists mostly of H_2 molecules with some D_2 and HD molecules. A third isotope, artificial and radioactive (by β–decay), has a mass number = 3. Designated 3 H, this isotope is known as *tritium* and has 2 neutrons and I proton in the nucleus. While diffusion and thermal diffusion methods may be used to separate the isotopes, the technical production of water rich in deuterium, heavy water, D_2O and HOD, is accomplished by electrolysis in Norway where there is abundant cheap hydroelectric power.

The H/D ratio in the gas discharged is less than in water, causing enrichment of the electrolyte liquor. Using about 2250 dm^3 of water from commercial electrolysis cells (already enriched), Lewis obtained 83 cm^3 of 99% pure D_2O. Deuterium and hydrogen (protium) will differ only in properties dependent on mass, enabling diffusion methods to be used in their separation. Chemically, the relatively large difference in mass (D = 2, H = 1) leads to a noticeable difference in reaction rates, a heavier atom being slower than a lighter atom with the same kinetic energy.

Farkas found that chlorine in the light at 30°C combines with hydrogen at three times the rate of its reaction with deuterium. The melting-point of deuterium, D_2, is 4.70°C higher and the boiling-point, 3.2°C higher than for hydrogen, H_2. The maximum density for heavy water, D_2O, deuterium oxide occurs at 11.22°C compared with 4.08°C for H_2O; furthermore, D_2O has a freezing-point 3.8°C higher and a boiling-point 1.4°C higher than water.

Hydrogen in combination

Period 1 of the Periodic Classification starts with hydrogen and is completed by the inert gas, helium, mass number = 4 and atomic number = 2, having a nucleus composed of two protons and two neutrons, with two external valency electrons. An atom of hydrogen may attain the inert gas electronic configuration of helium in two ways: either by sharing a pair of electrons to form a covalent compound or by acquiring another electron to form an ionic compound. Compounds in which hydrogen is united with one other element, binary compounds, are called *hydrides.* Consider examples: note that the valency of hydrogen cannot exceed unity because the valency shell cannot hold more than two electrons.

The hydrogen molecule, H_2

A covalent structure is in agreement with its properties. In passing, it may be mentioned that

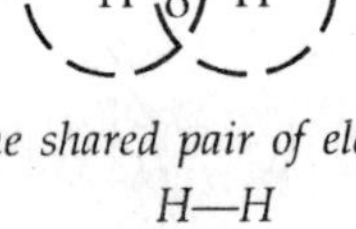

One shared pair of electrons

H—H

The two forms are in equilibrium, a fact not simply explained. At low temperatures para-hydrogen predominates but forms ortho-hydrogen as the temperature is raised until the limit ortho/para = 3/1 has been reached. The subject is too advanced for treatment here.

The hydrides

These are formed by sharing electrons. Simple hydrides of carbon, nitrogen, oxygen and fluorine provide examples. Although electrons are indistinguishable it helps to indicate their different origins in simple bond diagrams.

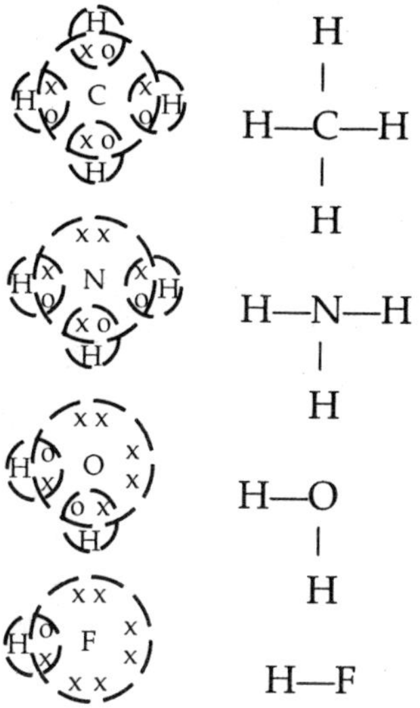

Consisting of molecules, covalent hydrides are colourless, electrically non-conducting compounds of low boiling-point. They are formed by non metals, except inert gases, and by some weak metals in which the tendency to form cations is not strong. Where chains of atoms are formed by an element, more than one hydride is possible:

$$CH_4, C_2H_6, C_3H_8, \text{etc. } NH_3, N_2H_4, HN_3$$

$$H_2O, H_2O_2$$

Ionic (salt-like) hydrides

in reactions with certain strong metals, which readily form cations, a hydrogen atom receives an electron to form the negative hydrogen ion, the hydride ion, H .

Lithium and calcium react:

$$[Li]^+ [Hox]^- \ LiH$$

$$[Ca]^{2+} [Hxo]_2^- \ CaH_2$$

Atomic number	Electronic configuration	
	Before	*After*
1	H 1	2 (He)
6	C 2, 4	2, 8
7	N 2, 5	2, 8
8	O 2, 6	2, 8 (Ne)
9	F 2, 7	2, 8

Atomic number	Electronic configuration		
	Before	*After*	
1	H 1	2	(He)
3	Li2, 1	2	(He)
20	Ca 2, 8, 8, 2	2, 8, 8 (Ar)	

These hydrides are colourless solids, of low volatility and with ionic lattices. They are known as the saline, or salt-like, hydrides. Although fusion brings slow decomposition, rapid electrolysis yields hydrogen at the anode in accordance with Faraday's Laws. The metal appears at the cathode. A few reactions may be mentioned, using lithium hydride and writing ionic equations. With water, lithium hydroxide and hydrogen are formed:

$$H^- + H_2O \rightarrow H_2 + OH^-$$

Anhydrous ammonia yields the amide:

$$H^- + NH_3 \rightarrow H_2 + NH_2^-$$

Carbon dioxide gives the methanoate (formate), $HCOO^-Li^+$; carbon monoxide, the methanoate and carbon:

$$H^- + 2CO \rightarrow H.COO^- + C$$

and sulphur dioxide, the dithionite,

$$2H^- + 2SO_2 \rightarrow S2O_4^{2-} + H_2$$

Interstitial (metal-like) hydrides

Hydrogen is occluded in non-stoichiometric proportions by certain transition metals. The compounds are examples of *berthollides.* Atoms of the gas penetrate into the layers of metal atoms (or ions) in the lattice. The strength of the bonding forces indicates a part chemical and part physical nature. Being metal-like, they are also called *the solid alloy type of hydride.* The final composition depends on the physical conditions and the state of division of the metal.

Other hydrides

Certain non-transition metals form non-volatile hydrides which are not ionic and are relatively unstable. Examples are BeH_2, MgH_2, AlH_3, ZnH_2, CdH_2, HgH_2, CuH.

Complex hydrides

Aluminium and boron form complex hydrides of some importance in which the Group III element is in the anion.

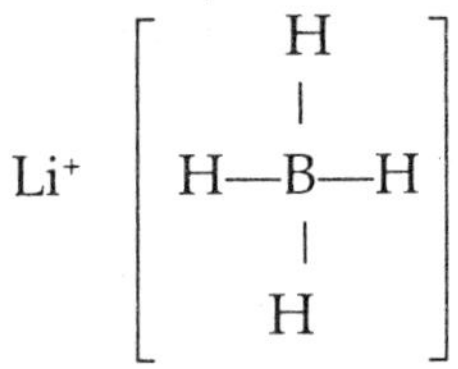

Lithium borohydride
Lithium tetrahydrido borate(III).

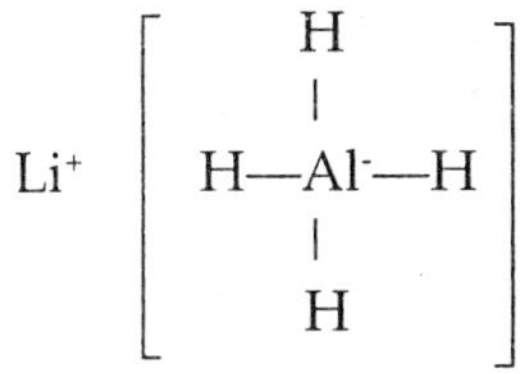

Lithium aluminium hydride
Lithium tetrahydrido aluminate(III)

Hydride type and the Periodic Table

A metal is an element which forms simple positive ions (cations) by the loss of electrons. The metals are described in a sort of verbal shorthand as strong or weak according to whether this tendency to form cations is also strong or weak. However, it is difficult to decide whether some elements are best classed as metals or non-metals. The strong metals form ionic hydrides. They are metal elements with low electronegativity (highly electropositive) found (and the same amount of hydroxide ions) but on in Groups I and II. Beryllium and magnesium, the dissolving an acid the concentration increases less electropositive members of Group II, form hydrides which are not of this simple type.

	I	*II*
Metals forming	Li	
ionic hydrides	Na	Ca
	K	Sr
	Cs	Ba

Calcium hydride is the least ionic. Hydrides of metals below sodium in Group I are highly reactive. Non-metals form covalent hydrides. But so do certain metals and metalloids. The stability of the hydrides falls, off with increase in atomic number. In Group V, BiH_3 was first detected as a radioactive gas in small yield from the action of acid on an alloy of magnesium and a radioactive bismuth isotope. This hydride must be placed in the progression

$$NH_3,\ PH_3,\ AsH_3,\ SbH_3 \text{ and } BiH_3,$$

where the hydrides prove progressively less stable with the development of metallic characteristics.

	III	*IV*	*V*	*VI*	*VII*
Elements forming	B	C	N	O	F
covalent hydrides		Si	P	S	Cl
		Ge	As	Se	Br
		Sn	Sb	Te	I
		Pb	Bi	Po	At

The hydrogen cation, H^+

The bare hydrogen nucleus, a proton of diameter about 10^{-15}m (1fm), as compared with 10^{-10}m (100pm) for other atoms, is much too small a point charge to exist freely in the presence of other atoms, except in a suitable discharge tube. In aqueous solution, it attaches itself to a water molecule to form the oxonium, hydroxonium or hydronium ion.

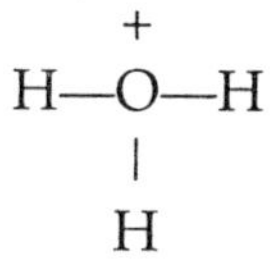

Acidity is ascribed to their presence. Pure water contains only 1×10^{-7} mole dm $^{-3}$ of these ions but on dissolving an acid the concentration increases (while the concentration of hydroxide ions decreases in accordance with the Equilibrium Law,

$$HCl + H_2O \rightleftharpoons H_3O+ + Cl^-$$

strong acid *ionization almost complete or complete*

$$HCN + H_2O \rightleftharpoons H_3O^+ + CN-$$

weak acid *ionization poor*

These equations are usually written simply:

$$HCl \rightleftharpoons H+ + Cl^-$$

$$HCN \rightleftharpoons H^+ + CN^-$$

where it is understood that the simple hydrogen ion is actually hydrated as $H+(H_2O)$ or possibly, as $H+(H_2O)2$. Usually the hydrated ion is taken as $H+(H_2O)$ or H_3O^+ and strictly, the name oxonium refers to this structure alone.

The hydrogen bond

The melting-points and boiling-points of the hydrides of

nitrogen, oxygen and fluorine are much higher than would reasonably be expected by comparison with corresponding data for hydrides of other elements in the respective groups.

The melting-point and boiling-point of methane, CH_4, follows the general trend of values for Group IV hydrides while abnormally high values appear for the first members of the other groups. This is explained by hydrogen bond formation. Because of its comparatively minute size and the inadequate screening of the nucleus by the shared pair of electrons, the combined hydrogen atom exerts a strong polarizing effect on other atoms. Hydrogen bonding is formed by compounds of the most electronegative elements, fluorine, oxygen and nitrogen.

The bond formed is electrostatic in origin and involves the hydrogen atom and an unshared electron pair. Because of the relatively large size of the atoms bridged and their mutual clectrostatic repulsion, only two atoms are linked by a hydrogen atom in this way; one covalently, the other electrostatically. This type of linkage can neither be described nor depicted simply.

The aggregates of molecules have reduced volatility and therefore higher boiling-points, stronger bonding forces and higher melting-points. Hydrogen fluoride Lends to form polymers with the angle, H—F. . . H = 140°. $(HF)_6$ is a very stable molecule of zig-zag structure.

The action of metals on water

Very electropositive metals liberate hydrogen from cold water, others require hot water or steam. In general, the vigour of action is greater the higher a metal is placed in the electrochemical series. The alkali metals (Group I) react vigorously with cold water. Less dense than water, globules of metal rush across the surface. The reactions are exothermic, the products being metal hydroxide solution and hydrogen.

The reactions increase in violence with increasing atomic number of the metal. Potassium is extremely violent: the gas ignites and burns with a flame coloured lilac by potassium vapour. Sodium does not ignite the hydrogen unless motion is restricted by placing the metal on a floating filter-paper when the hydrogen burns with a flame of yellow due to the sodium vapour. Lithium is brisk in action. Taking this as the example: lithium hydroxide solution remains,

$$2Li + 2H_2O \rightarrow 2LiOH + H_2$$

Various devices, none completely safe, have been devised for collecting hydrogen from the sodium reaction. For lithium, submerging the metal in a box made from gauze and holding this with tongs under a gas jar full of water, inverted in a trough of water, is probably the simplest way of collecting the hydrogen evolved. Metals forming sparingly soluble hydroxides or with a tenacious film of adhering oxide, may react slowly. Calcium sinks and reacts quietly with cold water, the slightly soluble hydroxide accumulating,

$$Ca + 2H_2O \rightarrow Ca(OH)_2 + H_2$$

Strontium and barium are increasingly vigorous, in that order. Magnesium decomposes boiling water slowly.

The action of metals on steam

Magnesium burns brilliantly when heated in steam, and leaves the oxide:

$$Mg + H_2O \rightarrow MgO + H_2$$

Zinc behaves similarly, but iron forms tri-iron tetroxide by -a reversible process:

$$3Fe + 4H_sO \rightleftharpoons Fe_3O_4 + 4H_2$$

Manganese, cobalt, nickel and tin liberate hydrogen at very high temperatures.

The action of metals on acids in solution

Metals above hydrogen in the electrochemical series liberate it from acids. The rate and extent of the reaction is determined by the state of purity of the metal, the solubility or insolubility of the product and the nature of the acid. An oxidizing acid will liberate hydrogen but rarely, as, for example, very dilute (1-2%) nitric acid with magnesium. The acids must be dissolved in water, because the process depends on the presence of hydrogen (oxonium) ions:

$$2H_3O^{+} + 2e^{-} - H_2 + 2H_2O$$

Except for aluminium, metals above tin in the electrochemical series liberate hydrogen from dilute acids. Aluminium and tin react with fairly concentrated hydrochloric acid. Very electropositive metals react violently. Magnesium and zinc are often used in the laboratory, the latter for gas in quantity. Neglecting hydration of ions,

$$Mg + 2H^{+} \rightarrow Mg^{2+} + H_2$$

e.g.

$$Mg + H_2SO_4 \rightarrow \underset{\textit{magnesium sulphate}}{MgSO_4} + H_2$$

Zinc reacts very slowly when pure but rapidly when impure, owing to galvanic action:

$$Zn + 2H^{+} - Zn^{2+} + H_2$$

Transfer of electrons is effected: zinc is oxidized and hydrogen ions reduced. Hydrogen may be dried over granular calcium chloride and collected over mercury. It reacts very slightly with concentrated sulphuric acid,

$$H_2SO_4 + H_2 \rightarrow SO_2 + 2H_2O$$

Hydrogen is usually obtained from a cylinder when required in quantity. However, a Kipp's apparatus may be used. Such an apparatus, charged prior to use and during use, is shown in Fig.1.

The action of certain metals on alkalis

Hot concentrated solutions of sodium hydroxide will react with aluminium and zinc to form respectively solutions of sodium aluminate and sodium zincate, with

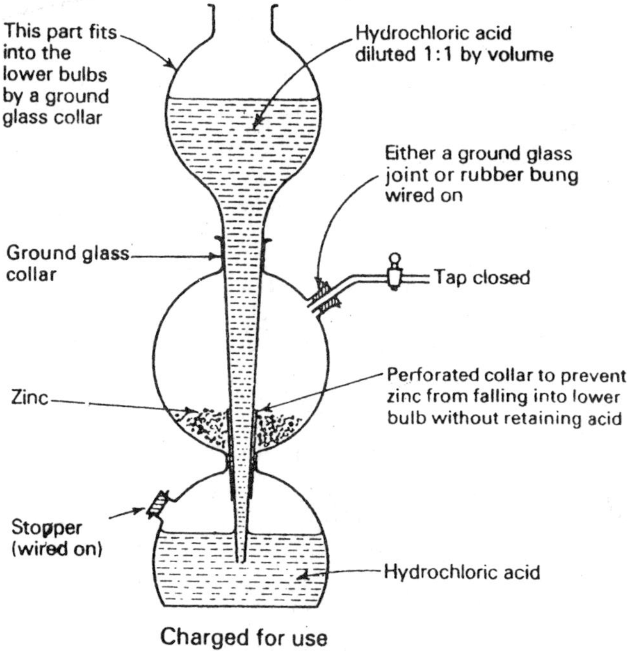

Fig. 1: Kipp's apparatus charged to deliver hydrogen

evolution of hydrogen. The aluminate and zincate ions may be shown in either the complex or the simplest form:

$$2\ Al + 2OH^- + 6H_2O \rightarrow \underset{\textit{aluminate ion}}{2Al(OH)_4^-} + 3H_2$$

or

$$2Al + 2OH^- + 2H2O \rightarrow \underset{\textit{aluminate ion}}{2AlO_2^-} + 3H_2$$

$$Zn + 2OH^- + 2H_2O \rightarrow \underset{\textit{zincate ion}}{Zn(OH)_4^{2-}} + H_2$$

or

$$Zn + 2OH^- \rightarrow \underset{\textit{zincate ion}}{ZnO_2^{\ 2-}} + H_2$$

Silicon gives sodium silicate while tin gives a mixture of sodium stannate(iv) and sodium stannate(iv). Other alkalis behave similarly.

The action of water on salt-like hydrides

Hydrolysis of ionic hydrides yields hydrogen and the alkali; essentially,

$$H^- + H_2O \rightarrow OH^- + H_2$$

Calcium hydride is produced commercially as a portable source of hydrogen, where the use of cylinders is not convenient, under the name hydrolith:

$$CaH_2 + 2H_2O \rightarrow Ca(OH)_2 + 2H_2$$

Electrolysis

Pure water is a poor conductor of electricity. However, aqueous solutions of acids, alkalis and certain salts may be used to prepare hydrogen by electrolysis. For example, electrolysis of separate aqueous solutions of sulphuric acid, sodium hydroxide and sodium sulphate yields two volumes of hydrogen at the cathode for every one volume of oxygen at the anode, when the electrolyte is saturated with gas. At the cathode, the net process is:

$$2H_3O^+ + 2e^- \rightarrow 2H_2O + H_2$$

For very accurate quantitative work with hydrogen, the purest gas is obtained by the electrolysis of warm baryta (barium hydroxide) solution between nickel electrodes. Contamination of the hydrogen by carbon dioxide is avoided because the latter is precipitated as barium carbonate. At the anode, oxygen is evolved:

$$OH^- \rightarrow OH + e^- \rightarrow 2OH \rightarrow H_2O + O$$

$$O + O \rightarrow O_2$$

Some diffusion of oxygen through the electrolyte occurs but contamination of the desired product, pure hydrogen, is minimized by the use of a U-tube for electrolysis. Hydrogen is purified by passage through a red-hot platinum gauze, which catalyses the combustion of traces of oxygen to water, absorbed in tubes containing solid caustic potash and phosphorus pentoxide.

The gas is stored by occlusion in palladium and may be regenerated on warming. Commercially, hydrogen is a product of the electrolysis of brine. In Tasmania, hydrogen is produced by the electrolysis of 28% aqueous caustic potash in mild steel cells, with mild steel sheet electrodes and nickel-plated anodes separated by asbestos cloth diaphragms. The cells are water-cooled. Distilled water is added to replace that decomposed.

The industrial production of hydrogen

Electrolytic methods have been described. 90% of all hydrogen made came from the water gas reaction. However, the pressure steam reforming of naphtha, the use of natural gas, mostly hydrogen and methane, and oil- refinery gas have come into prominence. In the USA, the breakdown for hydrogen production was natural gas (72%), refinery gas (11%), water gas (9%), electrolysis (6%) and coke oven gas (2%). Italy, France, Yugoslavia and Pakistan as well as the UK, also use natural gas. Middle East countries and the USA utilize oil-refinery gas. In the steel-producing regions of the UK, Germany, France and Belgium, coke oven gas, which contains 50-60% hydrogen, is important.

The reactions of hydrogen

Although comparatively inert at ordinary temperatures, large quantities of energy are liberated when hydrogen reacts with very electronegative elements. It reacts with fluorine in the dark and with chlorine in sunlight. Catalysis plays an important part in the reactions of hydrogen.

With elements; the synthesis of hydrides

Groups I and II

Except for beryllium and magnesium, the non-ionic hydrides of which are made indirectly, ionic hydrides 'are made by heating the metal in hydrogen. Unusual as it -may seem, the metal undergoes oxidation by loss of electrons while hydrogen is reduced:

$$2Na + H_2 \rightarrow 2NaH$$

$$Ba + H_2 \rightarrow BaH_2$$

Group III

No direct action.

Group IV

When an electric are is struck between carbon poles in an atmosphere of hydrogen, some hydrides of carbon are produced. The temperature is not less than 3000°C. Ethyne (acetylene) is the main product:

$$2C + H_2 \rightarrow C_2H_2$$

Group V

With a suitable catalyst, such as finely divided iron in the Haber process, hydrogen combines reversibly with nitrogen to form ammonia. The synthesis is exothermic. The physical conditions chosen vary but a temperature within the range 350-600°C and a pressure of 15-35 MPa (c. 150-350 atmosphere) may be used. The yield is 10-20%:

$$N_2 + 3H_2 \rightleftharpoons 2NH_3$$

Group V1

Hydrogen burns quietly in air with a very pale blue flame but in suitable proportions they form highly explosive mixtures. The product is water with traces of hydrogen peroxide. A more vigorous reaction is obtained with pure oxygen. Finely divided platinum catalyses the synthesis:

$$2H_2 + O_2 \rightarrow 2H_2O$$

In the oxy-hydrogen blow-pipe, the gases are delivered in concentric tubes, hydrogen burning in an atmosphere of oxygen. The temperature is about 2000°C.

When passed through boiling sulphur, hydrogen reacts to form some hydrogen sulphide:

$$H2 + S \rightleftharpoons H_2S$$

Selenium and tellurium react similarly.

Group VII

The vigour of reaction decreases with increase in atomic number of the halogen. Hydrogen inflames with fluorine, even at -250°C in the dark, while chlorine explodes in sunlight or ultraviolet light from a lamp or in the light of burning magnesium. However, chlorine and hydrogen react quietly in diffuse daylight or in the presence of activated charcoal or platinum as catalyst. Bromine and hydrogen react over a heated platinum catalyst. Hydrogen and iodine react reversibly on heating. Each halogen undergoes covalent reduction:

$$H_2 + F_2 \rightarrow \underset{\textit{hydrogen fluoride}}{2HF}$$

$$H_2 + Cl_2 \rightarrow \underset{\textit{hydrogen chloride}}{2HCl}$$

$$H_2 + Br_2 \rightarrow \underset{\textit{hydrogen bromide}}{2HBr}$$

$$H_2 + I_2 \rightleftharpoons \underset{\textit{hydrogen iodide}}{2HI}$$

Hydrogen chloride is synthesized commercially by burning chlorine in hydrogen, both gases being products of the electrolysis of brine.

Transition metals

Interstitial hydrides have been mentioned. In accurate quantitative work with hydrogen, the gas may be stored in palladium.

Addition to unsaturated compounds

This is described in Organic texts. Compounds containing double or triple bonds are said to be *unsaturated* and *react by addition.* Hydrogen usually reacts in the presence of heated finely divided nickel or colloidal palladium at room temperature. Examples include the reduction of ethane (an olefine) to ethane (a paraffin hydrocarbon), cyanomethane (a nitrile) to ethylamine (a primary amine), ethanal (an aldehyde) to ethanol (a primary alcohol) and propanone (a ketone) to propan-2-ol (a secondary alcohol). The reactions are general. Industrial examples include the reduction of carbon monoxide to methanol and the manufacture of margarine and cooking fat.

Reduction of oxides

Metals from iron downwards in the activity series may be obtained by reduction of their heated oxides in hydrogen:

$$\underset{\textit{lead monoxide}}{PbO} + H_2 \rightarrow Pb + H_2O$$

Monatomic hydrogen

Langmuir discovered that hydrogen, exposed to very low pressures of the order of 1 Pa (i.e. less than 0.01 mmHg) and heated to high temperatures by a tungsten, platinum or palladium wire carrying a heavy electric current, dissociated into atoms. Passage of molecular hydrogen at atmospheric pressure through an electric arc struck between tungsten electrodes also brings about dissociation:

$$H_2 \rightleftharpoons 2H; \quad \Delta H = +\ 432 kJ\ mol^{-1}$$

Experimental data are listed in Table 2. Atomic hydrogen is very reactive, forming hydrides with metals and non-metals at room temperature. It is, of course, a powerful reducing agent. At first sight it would seem that immediate re-combination of the atoms in atomic hydrogen would be likely to occur. However, a molecule formed by the collision of two atoms has enough energy to dissociate immediately. The surplus energy must be removed by 'three body' collisions, either with a third molecule, which will be rare, or with the surface of the containing vessel. Low pressures favour the survival of atomic hydrogen. At 67 Pa (0.5 mmHg) pressure, the 'half-life' of atomic hydrogen is 1/3 second. The great heat of re-combination may be used in welding. In the Atomic Hydrogen Blow-pipe or Welding Torch, a jet of hydrogen passes through an electric arc struck between tungsten poles and impinges upon the surface to be heated.

Table 2

The thermal dissociation of hydrogen H_2 2H

Temperature/K	*2000*	*2500*	*3100*	*3500*	*4000*
Percentage decomposition					
at 101 kPa	0.33	3.1	17	34	6.1
at 133 Pa	8.7	57.5	96	99.3	99.9

The temperature is above 3000°C. Although the hydrogen burns in the atmosphere, this is a secondary effect. Melting and welding is accomplished without surface oxidation in the reducing atmosphere of atomic hydrogen. Tungsten (m.p. 3400°C), tantalum (m.p. 2996°C) and thorium(iv) oxide (thorium dioxide) (m.p. 1825°C) readily melt with this technique.

Hydrogen oxide: water

Ice, water and steam are the solid, liquid and gaseous

forms of hydrogen oxide, H_2O. Water and ice, which are normally colourless, appear to have a bluish colour when viewed in depth. The melting-point of ice and the boiling-point of water under normal atmospheric pressure are taken as fixed points in thermometry, as 0° and 100° respectively, on the Celsius (Centigrade) scale. The critical temperature of water vapour is 374°C and the critical pressure, 22 MPa (218 atmosphere).

At 4°C and normal atmospheric pressure, the density of water is at a maximum. By definition (1901), the volume of I kilogram of water became the standard litre under these conditions, so that the density of water, by definition, was 1g ml^{-1} at 4°C. Scientists use internationally agreed units: those used in this book are recommended by the International Union of Pure and Applied Chemistry. By a resolution of the 12th Conference Generale des Poids et Mesures (CGPM) the term litre is now recognized as a special name for the cubic decimetre, but is not used to express high precision measurements.

The density of water at 4°C is now 1gcm^{-3}. The definition still applies in the UK for the purposes of the Weights and Measures Act. When the Act came into force on the yard was redefined as 0.9144 m exactly and the pound as 0.45359237 kg exactly. Water contracts when heated from 0°C to 4°C and expands on heating above 4°C. When lake water is cooled from ordinary temperatures, by the cold air above, the surface layer becomes denser and sinks.

A circulation of water is maintained until VC is reached. Further surface cooling produces water which is of a lower density and this remains in the surface until it freezes. If this did not occur, water would freeze from the bottom upwards, as happens in fast-moving steams where the water is thoroughly mixed. Sea water, with a relatively high proportion of dissolved solids, contracts continuously as the temperature falls. The density of ice is 0.917 gm^{-3}. Freezing is accompanied by expansion; water pipes which are split during a 'freeze-up' appear to 'burst' during the thaw.

The disintegration of rocks is brought about by the pressure of ice forming in the cracks and fissures. Plant cells are burst by the freezing of their contents. The triple-point of water is 0.0073°C at 624 Pa (4.68 mm Hg). Water is a very weak electrolyte and is almost electrically non-conducting at 25°C, the electric conductance is 4.0 μSm^{-1}. It is a highly associated liquid, with a high dielectric constant (= 8 1). The thermal dissociation of steam, imperceptible at 1000°C, is only 4% at 2300°C and 101 kPa pressure, hydrogen and oxygen showing a marked affinity for each other:

$$2H_2 + O_2 \rightarrow -2H_2O; \quad \Delta H = -2 \times 243 kJ\ mol^{-1}$$

The liberation of hydrogen from water may be achieved by the most electropositive metals (e.g. sodium) and from steam by a wider range of metals at higher temperatures.

The formula of steam

Early experiments on the synthesis of water were performed by Macquer and Priestley and Warltire. Cavendish determined the combining ratio of hydrogen to oxygen by volume, which he found to be 2.01: 1. Lavoisier realized the significance of these experiments in relation to his ideas on combustion.

Nicholson and Carlisle showed that water could be decomposed by electrolysis into hydrogen and oxygen. Davy proved that no other substance was formed and that the volume relationship, hydrogen: oxygen is 2 : 1. Gay-Lussac and von Humboldt determined the volume combining ratio of hydrogen and oxygen in twenty-four experiments; the average value of the ratio was

199.89 : 100 which they assumed equalled 200 : 100, allowing for experimental error. Further experiments by Bunsen confirmed this. In the laboratory demonstration, the combining ratio of hydrogen and oxygen by volume is determined by explosion in a eudiometer tube in which the gases are confined over dry mercury.

In some determinations an excess of hydrogen may be used, the excess being measured afterwards. But the electrolysis of dilute sodium hydroxide solution between nickel electrodes yields a pure gas with hydrogen and oxygen, from cathode and anode respectively, in the volume ratio of 2 : 1. This is dried by calcium chloride and introduced into the eudiometer of which several patterns are available. That of A. W. Hofmann is illustrated in Fig. 2.

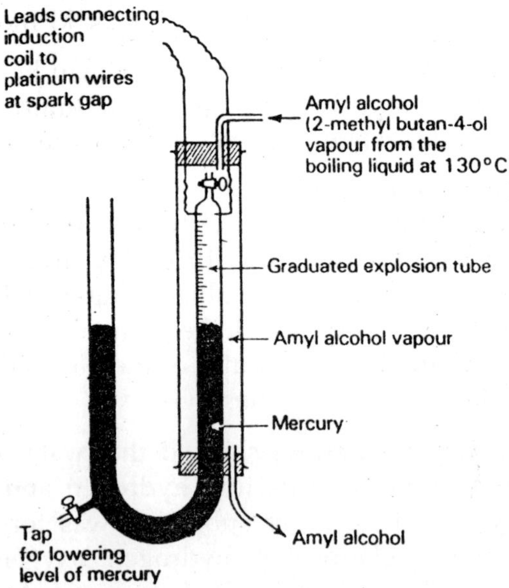

Fig. 2: Eudiometer for demonstrating the volume composition of steam.

With the heating jacket removed, mercury is introduced

until the whole of the eudiometer is full. By connecting a reservoir of dry electrolytic gas at the upper tap which is opened and running mercury away at the lower tap, the required quantity of gas is introduced into the graduated portion of the explosion tube. About 30 cm^3 is usually taken. The gas is now heated in the assembled jacket and the levels of mercury equalized.

When conditions are steady the volume of gas is noted or alternatively, the level marked on the jacket with a rubber band or gummed paper. Then mercury is run out of the lower tap until it occupies only about 150 mm of the tube above the tap and the open tube is corked to prevent loss of mercury during the explosion.

The cork is held and with the apparatus screened, a spark is passed. After levelling the mercury, the volume of residual gas is noted. The steam occupies two-thirds of the original volume. On cooling, this contracts to a negligible volume of water.

Hence, by Avogadro's Hypothesis,

$$2H_2 + O_2 \rightarrow 2H_xO_y$$

when it it seen,

$$x = 2,\ y = 1$$

$$\therefore \text{ formula of steam is } H_2O$$

Burt and Edgar have shown, as the result of 59 determinations, that the volume combining ratio for hydrogen and oxygen is 200.288 : 100. The ratio is not 200.000 : 100 because neither gas obeys the ideal gas equation exactly. Oxygen and hydrogen were made by electrolysis of warm baryta solution, purified very carefully and dried. To avoid temperature and pressure corrections, all measurements were made at 0°C and 101 kPa (760 mmHg) pressure.

The gravimetric composition of water

Dumas determined the combining ratio of hydrogen and oxygen by mass. An excess of purified dry hydrogen was passed over a weighed quantity of heated dry copper (ii) oxide and the water collected by absorption in calcium chloride, solid potassium hydroxide and sulphuric acid or phosphorus pentoxide.

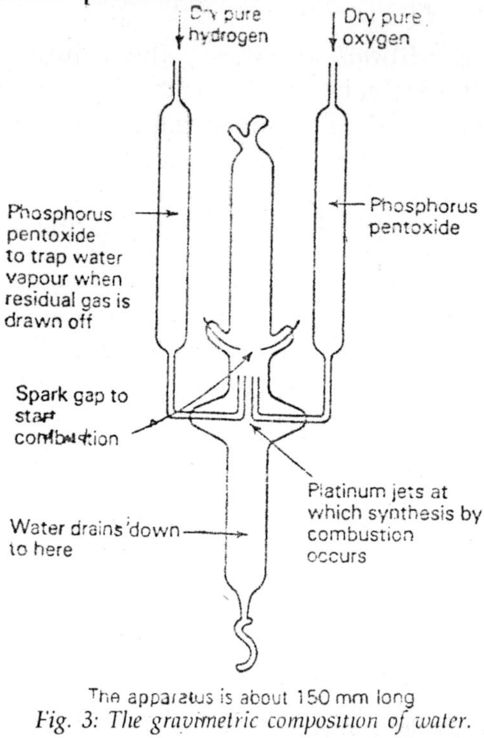

The apparatus is about 150 mm long

Fig. 3: The gravimetric composition of water.

The loss in mass of the copper (ii) oxide gave the mass of oxygen which forms water, the mass of which was also measured. Thus, the ratio in which hydrogen and oxygen combine by mass could be calculated:

Mass of hydrogen : mass of oxygen = 1.002: 8

Morley in employed direct synthesis; his apparatus is shown in Fig. 3. The apparatus is evacuated and weighed. Oxygen and hydrogen, purified and collected in weighed globes, are introduced separately as shown and burned at platinum jets, action being started by electric sparking across the spark gap.

In later experiments, hydrogen was stored in palladium from which it could be expelled by heating. After the experiment, the water is frozen and the residual gas (oxygen) pumped out, any water vapour being trapped by phosphorus pentoxide. The gas is weighed and analysed. The apparatus isre-weighed. In this method the separate masses of oxygen and hydrogen are measured and also the mass of water formed. This provides a check: the sum of the masses of oxygen and hydrogen must equal the mass of water. About 30 g of water were synthesized in each experiment.

Morley found, as the average of a series of 9 experiments, that the combining ratio of hydrogen to oxygen is 1.00768 : 8 (exactly). Using Morley's values of the normal densities of oxygen and hydrogen, Burt and Edgar (1916) calculated, from their volumetric composition results, the gravimetric composition of water. The combining ratio of hydrogen and oxygen by mass came to 1.00773 : 8.

Water supplies

Water occurs abundantly in the oceans, lakes and rivers, the atmosphere and as water of hydration in rocks. Sea water contains about 3.6% of dissolved solids by weight, including about 2.7% sodium chloride, common salt. Evaporation of sea water yields sodium chloride, magnesium chloride and sulphate, calcium sulphate and carbonate (from hydrogen carbonate), potassium chloride and bromide; the following ions are present: Na^+, K^+, Mg^{2+}, Ca^{2+}, Cl^-, Br^-, HCO_3, $SO_4{}^{2-}$ together with smaller quantities of other substances.

Of natural waters, rain water is the purest form. It is precipitated from water vapour which has evaporated from the seas and lakes. Away from towns, the impurities are the dissolved gases oxygen, nitrogen and carbon dioxide from the atmosphere, and nitrogen and ammonium salts from electrical action on air. Near industrial areas, sulphur dioxide, which forms sulphurous acid and then sulphuric acid, and soot are present.

The impurities gathered by stream, river and lake water depend very much on the rock formations over and through which the water has passed. They include chlorides, sulphates, nitrates, hydrogen carbonates and carbonates of sodium, magnesium, calcium and iron to which may be added animal and vegetable refuse, sewage and industrial effluents.

A highly industrialized society like ours depends to a marked degree on a piped supply of clean, safe water for domestic and commercial purpose. Water for consumption must be purified if disease and epidemics are to be avoided. Where water comes from a river like the Thames, it must be stored for weeks or months to allow suspended matter to settle and many harmful bacteria to die. Other sources, such as moorland catchment areas and limestone regions, yield water which can be processed immediately.

Coarse debris is removed by filtration through sand and gravel and pressure filtration through sand, coupled with the action of chemical coagulants. These include aluminium sulphate, sodium aluminate and silicate, which remove algae, bacteria and other fine suspended matter by absorption on precipitated aluminium hydroxide and hydrated silica. There is increasing concern over the pollution of water for drinking by the residues of fertilizers, insecticides and herbicides, used in intensive farming and from industrial effluent. The EEC has produced regulations requiring governments to adopt stringent purity specifications for water, especially over the permitted concentration of nitrate ion, difficult and expensive to remove.

The nitrite ion, NO_2, is suspected of being capable of causing cancer. Water catchment areas are sign posted as such in certain countries e.g. western Germany, a move which has much to commend it when backed up by legal requirements. There is no instant solution to the problem, or series of solutions, because the chemicals which cause the pollution take very many years to reach the levels of the underground water reservoirs.

Water for drinking is sterilized by chlorine in a concentration of 1-2 parts per million. While the water is now fit for drinking and domestic use, it may need further treatment for industrial purposes. Finally, let it be remembered that the fame of not a few beers rests firmly on the very high quality of the water from private wells or reservoirs associated with the brewery and this is used as a selling point e.g. Burton (UK), Bitburger (Germany). Paradoxically, purification measures enforced by law may result in some very pure waters becoming less so.

The habit of buying bottled water has spread across Europe from France. It is quite instructive to read the declared analysis of contents on the labels of such bottles, as much for what is stated as for what is not. It should also be remembered that while some brands are no doubt excellent, others have been sharply criticised by consumer associations. Whatever the merits or otherwise of what is written here, it is abundantly clear that EEC concern over the quality of water supplies is not misplaced.

Water pollution by fertilizers

Not all the crop nutrients applied as fertilizers are used

by the crops. The unused portions represent a financial loss to the farmer and an environmental hazard to us all. The entry of nitrate into drinking water will cause acute commercial and scientific problems for the immediate future.

The EEC nitrate target of less than 50 parts per million in drinking water will be extremely difficult to reach in areas of intensive farming such as East Anglia and Lincolnshire and there is informed debate about whether this target is too stringent.

The hardness of water

A hard water will not readily form a lather with soap. Soap solution forms stable bubbles of lather by lowering the surface tension. Soaps are sodium and potassium salts of long-chain fatty acids; e.g. sodium octadecanoate, sodium stearate , $C_{17}H_{35}COONa$. Hard water contains calcium or magnesium ions or both, as hydrogencarbonates, sulphates or chlorides.

The stearates of these metals are insoluble. On adding soap to hard water a precipitate of calcium octadecanoate or magnesium octadecanoate appears as a scum. Extra soap is required before the water has been softened:

$$Ca^{2+} + 2C_{17}H_{35}, COO^- \rightarrow (C_{17}H_{35}COO)_2Ca(s)$$

The presence of scum will spoil the finish on materials which have been laundered and on fabrics in the textile industry. The use of synthetic detergents avoids this. They have either soluble calcium and magnesium salts or are non-ionic, being described as synthetic to distinguish them from soaps which are made from vegetable and animal oils.

In the context of the domestic washing of clothes the concept of hardness seems distinctly old-fashioned. Sometimes hardness can be removed by boiling and this leads to the distinction between two types of hardness.

Temporary (or alkaline) hardness of water may be removed by boiling.

Permanent (or non-alkaline) hardness of water cannot be removed by boiling.

Temporary hardness is due to the hydrogen-carbonates of calcium and magnesium. They are formed when rain water containing dissolved carbon dioxide, some of it 'as carbonic acid, meets limestone, magnesite, dolomite or similar carbonate rocks of calcium and magnesium. The hydrogencarbonate ion is decomposed by heat and the metal carbonate is re-precipitated. For calcium hydrogencarbonate,

$$C_aCO_3 + H_2O + CO_2 \underset{heat}{\overset{formation}{\rightleftarrows}} Ca(HCO_3)_2$$

Not only is hardness wasteful of soap but it is a source of expense and some danger in industrial plant using boilers. The removal of temporary hardness by boiling leaves behind a boiler 'scale' of calcium and magnesium carbonates. Extra heat must now be supplied to heat the water within the barrier of scale and extra pumping power must be supplied, to drive water through congested pipes, which may become blocked.

Permanent hardness is caused by the chlorides and sulphates of magnesium and calcium. They accumulate in steam-raising plant so that eventually the water becomes saturated. Crystallization occurs and deposition on the inside of boilers begins. Problems of hardness affect the running of power stations, ships and hot-water systems. Minerals dissolved in natural waters are responsible for the distinctive 'Spa' waters, and probably for distinctive types of beer and drinks associated or with particular regions.

Hardness in water will seriously affect the processes in the dyestuff industry and in tanning leather. Removal of hardness in water is called, appropriately, softening.

The softening of water

Temporary hardness can be removed by boiling but this is expensive. The hydrogencarbonate ion is decomposed on heating,

$$2HCO_3^- \rightarrow H_2O + CO_2 + CO_3^{2-}$$

and precipitation of carbonate occurs,

$$Mg^{2+} + CO_3^{2-} \rightarrow MgCO_3(S)$$

$$Ca^{2+} + CO_3^{2-} \rightarrow CaCO_3(s)$$

Solutions of hydrogencarbonates are alkaline by salt hydrolysis, which explains the term 'alkaline hardness', although they are, in fact, acid salts. Removal of the hydrogen, the acid part, by alkali leaves carbonate ions which are precipitated as before. Addition of the calculated quantity of slaked lime will destroy temporary hardness,

$$HCO_3 + OH^- \rightarrow H_2O + CO_3^{2-}$$

$$Ca^{2+} + CO_3^{2-} \rightarrow CaCO_3(S)$$

Extra lime must be avoided because a solution of slaked lime, calcium hydroxide, shows permanent hardness. Both types of hardness usually occur together.

The degree of hardness may be determined by titration against standard soap solution, before and after boiling, to give the relative amounts of temporary and permanent hardness. Alkalinity and the concentration of other ions such as chloride and phosphate (used in water treatment) are also determined titrimetrically.

Softening by precipitation

The lime-soda method

Calculated quantities of lime (calcium hydroxide) and sodium carbonate are added; the former removes temporary hardness and the latter, permanent. In fact, sodium carbonate would remove both kinds of hardness but the use of lime is cheaper. The processes may be summarized by equations; for calcium:

$$HCO_3 + OH^- \rightarrow CO_3^{2-} + H_2O$$

$$Ca^{2+}\ CO_3^{2-} \rightarrow CaCO_3(s)$$

or

$$Ca(HCO_3)_2 + Ca(OH)_2 \rightarrow 2CaCO_3(s) + 2H_2O$$

permanent:

$$Ca^{2+} + CO_3^{\ 2-} \rightarrow CaCO_3(s)$$

$$CaSO_4 + Na_2CO_3 \rightarrow CaCO_3(s) + Na_2SO_4$$

$$CaCl_2 + Na_2CO_3 \rightarrow CaCO_3(s) + 2NaCl$$

Softening by precipitation

The use of caustic soda (and sodium carbonate)

Carefully controlled quantities of caustic soda are used to precipitate magnesium hydroxide and calcium carbonate from temporary hard water. Magnesium hydroxide is less soluble than the carbonate and is therefore precipitated. The full equations show that sodium carbonate solution is a product,

$$Ca(HCO_3)_2 + 2NaOH \rightarrow CaCO_3(s) + Na_2CO_3 + 2H_2O$$

$$Mg(HCO_3)_2 + 4NaOH \rightarrow Mg(OH)_2(S) + 2Na_2CO_3 + 2H_2O$$

Any permanent hardness may be destroyed by two methods. That due to magnesium by caustic soda,

$$Mg^{2+} + 2OH^- \rightarrow Mg(OH)_2(s)$$

and that due to calcium either by the sodium carbonate generated above or by this plus extra sodium carbonate added for the purpose,

$$Ca^{2+} + CO_3^{2-} \rightarrow CaCO_3(s)$$

If the softening treatment results in too much sodium carbonate being left in the water, the limesoda process is preferred. The coagulation of magnesium hydroxide is helped by the addition of sodium aluminate, the aluminium hydroxide formed aiding flocculation.

Exchange methods for softening

The zeolite softening process is frequently used in the home, in the textile industry and by some water authorities. Zeolite mineral is a complex sodium aluminium silicate and hard water flowing through it leaves behind calcium and magnesium ions, acquiring the equivalent quantity of sodium ions. This is also called the Base-Exchange method.

Regeneration of the zeolite is achieved by addition of strong salt solution, when the trapped magnesium and calcium ions pass into solution. While the process is expensive, it is useful where hardness is slight. The Hydrogen Ion Exchange process utilizes the carbonaceous product of the action of sulphuric acid on coal. Metal ions are replaced by hydrogen ions so the product is acidic and is neutralized by passage through another type of exchange material. By the use of special Ion-Exchange Resins both anions and cations can be removed from water and the product is comparable with distilled water.

Phosphate treatment

A highly polymerized phosphate, sodium metaphosphate, $(NaPO_3)_n$, with added alkali, is sold under the name 'Calgon' for water treatment. Its applications are many. Small quantities will *sequester* calcium and magnesium ions, removing the ions as complex phosphates. It is used as a domestic softening agent, to prevent scale formation in hot-water supply systems, to prevent corrosion of iron and steel water mains by soft moorland water and to hold in solution compounds of iron which would otherwise be precipitated by oxidation of iron(ii) hydrogencarbonate in well waters. In boiler water treatment, Calgon may be used to ensure controlled precipitation of phosphate and carbonate sludge without scale formation. It is used either by itself with soaps in textile scouring and laundry processing or in detergents marketed for dish washing and bottle washing. There are many other industrial applications.

Corrosion by soft waters

Iron and lead pipes in hard-water regions are rapidly coated by compounds formed by minerals in the water carried. Very soft water from moorland catchment areas will corrode pipes made from iron or lead especially if it is relatively acidic. Water is coloured red by iron compounds and lead causes poisoning.

The water is just alkaline to methyl orange. Increasing the pH by carefully controlled additions of lime, sodium carbonate and sodium hydroxide will make the water alkaline enough to prevent corrosion. Sodium silicate is particularly effective in preventing both types of corrosion. It provides, with the deposition of hydrated silica, a silicaceous lining to the pipes and unlike lime, does not introduce hardness.

Stalactites and stalagmites

When rain water, with its dissolved carbon dioxide, falls on to limestone rocks, calcium hydrogencarbonate solution is formed:

$$CaCO_3 + H_2O + CO_2 \rightleftharpoons Ca(HCO_3)_2$$

On reaching a cave, it drips from the roof. Decomposition occurs and calcium carbonate is precipitated. Hanging icicle-like deposits called stalactites are formed, growing at the rate of a fraction of an inch every year. On the floor of the cave, the continuous dripping leads to the upward growth of stalagmites. Wherever a stalactite and stalagmite meet, a pillar is formed. In the Mendip caves and elsewhere, stalactites, stalagmites and pillars of incredible beauty abound.

Reactions of water

The liberation of hydrogen by certain metals acting on water and steam has been described. Of the nonmetals, fluorine, chlorine, carbon and silica decompose water. Fluorine decomposes water with the formation of hydrofluoric acid and ozonized oxygen, while chlorine in sunlight produces hydrochloric acid and oxygen,

$$2F_2 + 2H_2O \rightarrow 4HF + O_2 \text{ (and } O_3\text{)}$$

$$2Cl_2 + 2H_2O \rightarrow 4HCl + O_2$$

Steam, passed over carbon at red to white heat, yields water gas,

$$C + H_2O \rightleftharpoons CO + H_2$$

while silicon very slowly forms silica and hydrogen,

$$Si + 2H_2O \rightarrow SiO_2 + 2H_2$$

With oxides

The oxides of certain strong metals i.e. those with a powerful tendency to form cations, react with water to form solutions of the hydroxide:

e.g. $$BaO + H_2O \rightarrow Ba(OH)_2$$

while oxides of many non-metals form acids:

e.g. $$SO_3 + H_2O \rightarrow H_2SO_4$$

Decomposition by water: hydrolysis

Hydrolysis is the decomposition of compounds by water. Where the balance between the hydrogen and hydroxide ion concentrations is disturbed with a resulting change in the pH, but without an observed decomposition, the term salt *hydrolysis is* applied. pH is a measure of the acidity:

$$pH = -\log_{10} [H^+]/\text{mol dm}^{-3}$$

where $[H^+]$, is the hydrogen ion concentration, mol dm^{-3} and the temperature is usually standardized at 25°C or 18°C.

Binary compounds of metals and non-metals may be hydrolysed to hydrides of the non-metal:

$$Al_4C_3 + 12H_2O \rightarrow 4Al(OH)_3 + 3CH_4(g)$$

aluminium carbide … *methane*

$$CaC_2 + 2H_2O \rightarrow Ca(OH)_2 + C_2H_2(g)$$

calcium carbide … *ethyne (acetylene)*

$$Mg_2Si + 4H_2O \rightarrow 2Mg(OH)_2 + SiH_4(g) \text{ etc.}$$

magnesium silicide … *silane*

$$Mg_3N_2 + 6H_2O \rightarrow 3Mg(OH)_2 + 2NH_3(g)$$

magnesium nitride … *ammonia*

$$Ca_3P_2 + 6H_2O \rightarrow 3Ca(OH)_2 + 2PH_3(g)$$

calcium phosphide … *phosphine*

$$MgS + 2H_2O \rightarrow Mg(OH)_2 + H_2S(g)$$

magnesium sulphide … *hydrogen sulphide*

Consider the hydrolysis of chlorides, taking the chlorides XCl_3 of the Group V elements nitrogen and phosphorus as examples. Nitrogen trichloride, a yellow oily liquid, is slowly hydrolysed and the products, ammonia and hypochlorous acid, other wise chloric(i) acid (shown in the equation), react to form nitrogen, nitric acid and hydrochloric acid:

$$NCl_3 + 3H_2O \rightarrow NH_3 + 3HOCl$$

Phosphorus trichloride, a liquid, is rapidly hydrolysed to phosphorous acid and hydrogen chloride:

$$PCl_3 + 3H_2O \rightarrow H_3PO_3 + 3HCl$$

Hydrolysis usually gives the hydroxy compound of the element and not a hydride as occurred with nitrogen trichloride. Thus, boron trichloride forms boric and hydrochloric acids,

$$BCl_3 + 3H_2O \rightarrow H_3BO_3 + 3HCl$$

The extent of hydrolysis is indicative of the nature of an element: metal or non-metal, weak or strong metal. Salts with oxoacids are formed only by metals, but those of weaker metals are readily hydrolysed or only exist as basic salts. Bismuth nitrate is hydrolysed to an oxonitrate or subnitrate, called bismuth oxide nitrate:

$$Bi(NO_3)_3 + H_2) \rightarrow BiONO_3 + 2HNO_3$$

Leas(rv) sulphate and acetate are instanly decomposed by water,

$$Pb(SO_4)_2 + 2H_2O \rightarrow PbO_2 + 2H_2SO_4$$

$$CH_3(COO)_4Pb + 2H_2O \rightarrow PbO_2 + 4CH_3COOH$$

lead(iv) oxide (lead dioxide) being precipitated.

Structure of water and hydration

The bond diagram for water may be written,

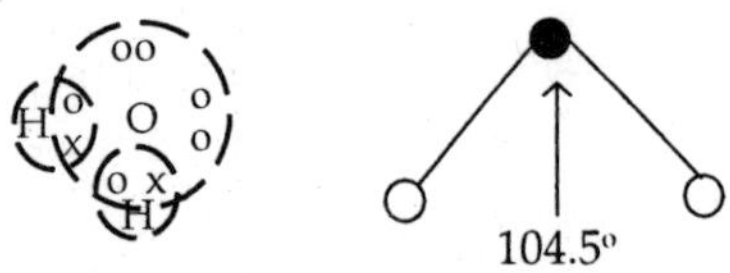

in which hydrogen and oxygen acquire the electronic configurations of helium and neon respectively. The structure of water has been investigated by infrared spectroscopy. The H—O internuclear distance is 0.0915 nm and the HOH = 104.5°. That the water molecule is not linear is supported by the fact that the dipole moment is 6.0×10^{-28}cm (= 1.81)) whereas for a linear molecule it would be zero. Oxygen with a high affinity

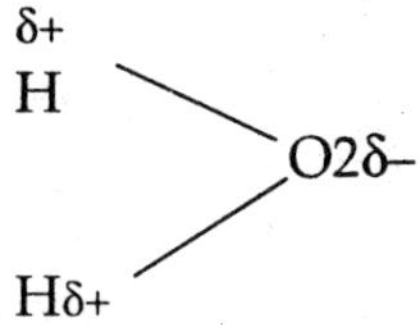

for electrons, has a residual ne-Lative charge while the hydrogen atoms are positive. This accounts for the association of water molecules with both anions and cations in solution and other water molecules.

The positive hydrogen is electrostatically attracted to anions and the negative oxygen to cations, although oxygen may enter into co-ionic bond formation as in the oxonium ion:

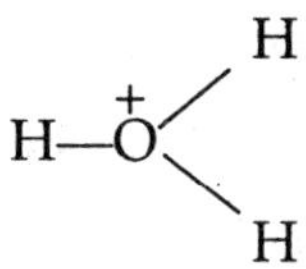

Without association of water molecules, the melting-point of water might reasonably be expected to be in the region of—100°C and the boiling-point about 80°C.

The hydrated cations of transition metals and copper(ii) have characteristic colours: Mn^{2+} (pink), Fe^{2+} (pale green), Ni^{2+} (deep green), CO^{2+} (deep pink) and Cu^{2+} (blue).

Water of crystallization

Cooling saturated aqueous solutions of many salts and certain acids and bases, yields crystals containing the compound associated with water in definite molecular proportions. The isolated crystals are perfectly dry. These additive molecular compounds, containing water of crystallization, are termed *hydrates* and may be easily decomposed by heat to give the anhydrous compound, although some salts undergo hydrolysis. Blue vitriol, copper(ii) sulphate pentahydrate, loses four-fifths of the water of hydration at about 100°C and the remainder above 200°C, leaving a white (microcrystalline) powder:

$$CUSO_4 . 5H_2O \rightarrow CUSO_4 . H_2O + 4H_2O(g)$$

$$CUSO_4 . H_2O \rightarrow CUSO_4 + H_2O(g)$$

Four molecules of water are associated with the copper(ii) ion and one with the anion, holding the crystal structure together. Magnesium chloride, $MgCl_{2,}$ $6H_2O$, and iron(iii) chloride, $FeCl_3.6H_2O$, yield the oxide and lose hydrogen, chloride and water on heating. Hydrates of acids and bases include ethanedioic acid (oxalic acid), $(COOH)_2.2H_2O$, and lithium hydroxide, LiOH. H_2O.

The zeolite minerals (complex aluminium silicates), used in water-softening, lose water without change of crystal structure. Sometimes, this water is called *zeolytic water* to distinguish it from water of crystallization.

Efflorescence and deliquescence

On exposure to the atmosphere certain hydrates lose water, or effloresce. Translucent crystals of sodium carbonate decahydrate (washing soda), Na_2CO_3 $10H_2O$, become coated with a powdery deposit of monohydrate, appearing white,

$$Na_2CO_3\ {}_{10}H_2O \rightleftharpoons Na_2CO_3 . H_2O + 9H_2O$$

Crystals of sodium sulphate decahydrate form the anhydrous salt,

$$Na_2SO_4 . 10H_2O \rightleftharpoons Na_2SO_4 + 10H_2O$$

For this to happen at a given temperature, the vapour pressure due to water produced by dissociation must exceed the partial pressure of water vapour in the atmosphere. This is usually about two-thirds of the saturated vapour pressure of water at ordinary temperatures, between 1.3 – 2.0 kPa (10-15 mmHg), while for sodium carbonate decahydrate and sodium sulphate decahydrate the dissociation pressures of water vapour are 3.23 and 3.71 kPa (24.2 and 27.8 mmHg) at 18°C respectively. Deliquescence occurs where a substance absorbs moisture from the atmosphere to form a solution. The saturated vapour pressure of water over the solution formed (saturated initially) is less than the vapour pressure due to water in the atmosphere at that temperature.

Absorption of water continues until they are equal. The vapour pressures at saturation due to water over saturated solutions of calcium chloride and iron(iii) chloride at 18°C are 427 and 800 Pa (3.2 and 6.0 nimHg) respectively. The hexahydrates of calcium and iron(iii) chloride are deliquescent. Where the water vapour pressure is less than that in the atmosphere and the water vapour pressure of the solution exceeds that of the atmosphere, hydrated compounds show neither efflorescence nor deliquescence.

Hygroscopic substances

Solids which absorb Moisture from the atmosphere without appearing wet, and liquids which absorb moisture, are termed *hygroscopic.* In quantitative work, copper(ii) oxide must be heated and cooled in a desiccator before weighing because it absorbs water. Sulphuric acid on the other hand is used for drying, gases being bubbled through the liquid. A desiccator is a closed vessel for storing a substance in the presence of a compound which absorbs moisture. Calcium chloride, sulphuric acid and silica gel are among the compounds chosen for their water absorption properties.

The detection and estimation of water

Water may be characterized by the determination of its melting- and boiling-points. A qualitative test for traces of water in liquids employs anhydrous copper(ii) sulphate which turns from white to the blue hydrate. In quantitative experiments, water may be absorbed in previously weighed tubes of phosphorus pentoxide or calcium chloride, suitably protected from water vapour in the atmosphere. Solids are heated to constant mass with cooling in a desiccator in the determination of water, if the compound is otherwise stable.

One volumetric method for estimating water in liquids requires the addition of calcium carbide, the volume of ethyne (acetylene) produced with any water present being measured,

$$CaC_2 + 2H_2O \rightarrow Ca(OH)_2 + C_2H_2(g)$$

Deuterium compounds

Deuterium, the naturally occurring isotope of hydrogen, with mass number = 2, due to the neutron acquired by the proton of the hydrogen nucleus.

Deuterium oxide, 'heavy water'

By prolonged electrolysis of water, in which the *Salts:* lighter hydrogen (protium) isotope is preferentially discharged, deuterium oxide accumulates. This was first accomplished by Washburn and Urey. Pure deuterium oxide was prepared by Lewis and Macdonald in the following year. They obtained 83cm^3 of 99% purity, using 2250dm^3 of residual water from commercial electrolysis cells. Heavy water contains both D_2O and HOD,

D_2O is available commercially and has been studied in detail. The melting-point is 3.8°C and the boiling-point, 101.4°C.

The preparation of deuterium compounds

Deuterium oxide may be used to prepare deuterium compounds. These have slightly different physical constants from the corresponding compounds containing the lighter isotope, but undergo the same chemical reactions. Although the chemical reactions are the same as with ordinary water, it is instructive to outline the synthesis of deuterium compounds starting from deuterium oxide. Compounds are named by introducing the prefix deutero before the more usual name or by displacing hydrogen by deuterium where appropriate.

ND_3, deuteroammonia:

$$Mg_3N_2 + 6D_2O \rightarrow 3Mg(OD)_2(s) + 2ND_3(g)$$

and

$$ND_3 + D_2O \rightleftharpoons ND_4^+ + OD^-$$

NaOD, sodium deuteroxide:

$$2Na + 2D_2O \rightarrow 2NaOD + D_2(g)$$

D_2, deuterium:

Deuterium may be prepared by the action of sodium on D_2O or by the electrolysis of D_2O containing alkali (e.g. NaOD, etc.) or acid (e.g. D_3PO_4, D_2SO_4, etc.) or an oxosalt (e.g. Na_2SO_4).

Deuterophosphoric acid may be prepared by the action of phosphorus pentoxide on deuterium oxide:

$$2D_2O + P_4O_{10} \rightarrow 4DPO_3$$

$$6D_2O + P_4O_{10} \rightarrow 4D_3PO_4$$

D_2SO_4, deuterosulphuric acid

$$SO_3 + D_2O \rightarrow D_2SO_4$$

and similarly, DNO_3 and $DClO_4$.

Salts

$$D_2SO_4 + 2ND_3 \rightarrow (ND_4)_2SO_4$$

$$D_2SO_4 + 2NaOD \rightarrow Na_2SO_4 + 2D_2O$$

$$Na_2SO_4 + {}_{10}D_2O \rightarrow Na_2SO_4 .\ 10DO$$

$$CuO + D_2SO_4 \rightarrow CUSO_4 + D_2O$$

$$CUSO_4 + 5D_2O \rightarrow CUSO_4.5D_2O$$

Insoluble deuteroxides:

$$Al^{3+} + 3OD^- \rightarrow Al(OD)3(s)$$

Halogen deuterides:

$$D_2(g) + Cl_2(g) \rightarrow 2DCl(g)$$

$$NaCl + D_2SO_4 \rightarrow NaDSO_4 + DCl(g)$$

$$P_4 + 10I_2 + 16D_2O \rightarrow 20DI(g) + 4D_3PO_4$$

Deuterocarbons: deuteroethyne (deuteroacetylene):

$$CaC_2 + 2D_2O \rightarrow C_2D_2(g) + Ca(OD)_2$$

C_2D_2 is the source of many organic compounds.

General properties of acids, bases and salts in relation to water

Acids

Acids belong to a group of compounds which have certain similar characteristics when dissolved in water. They taste sour, deriving their name from the Latin, *acidus* = sour. Acids corrode the more reactive metals liberating hydrogen, produce an effervescence of carbon dioxide with carbonates and change the colour of certain dye solutions, called, appropriately, indicators.

The vegetable dye, litmus, is turned from purple to red by acids. Acids were investigated by Boyle. The characteristic properties of acids are neutralized by reaction with compounds having another group of characteristics in common.

These compounds were named bases by Rouelle in and include oxides and hydroxides of metals. In a way, acids and bases are chemical opposites. When an acid in solution reacts with a metal, hydrogen is evolved. Where does it come from? The acid, the water or both? All acids may be shown to contain hydrogen, replaceable by a metal with the formation of a salt, but this does not distinguish acids alone. For example, caustic soda solution attacks aluminium with the liberation of hydrogen. Solutions of different acids are electrically conducting to varying degrees for corresponding concentrations and this is due to the dissociation of the acid molecule into charged ions. The appearance of 'hydrogen ions' accounts for the closely similar behaviour of acids in aqueous solution. An example will illustrate the role of the water: freshly prepared nitric acid, a liquid, is without action on marble chippings (calcium carbonate) but addition of a little water results in a vigorous reaction. The ionization of nitric acid should be written:

$$HNO_3 + H_2O \rightarrow H_3O^+ + NO_3^-$$

yet it is common practice to use the simpler equation of the Arrhenius Ionic Theory which may be more familiar:

$$HNO_3 \rightleftharpoons H^+ + NO_3^-$$

but it must be remembered that the presence of water is essential for the display of acidic behaviour and that the H^+ ion does not exist, as such, in aqueous solution. From the ionic standpoint for aqueous solutions an acid may be simply defined:

An *acid* is a compound which ionizes to produce hydrogen ions (strictly, oxonium ions) when dissolved in water. The oxonium ion is sometimes called the hydronium or hydroxonium ion. The term is analogous to ammonium. When there is potentially one hydrogen ion in each molecule the acid is monobasic (e.g. nitric acid), two ions, dibasic (e.g. sulphuric acid) and three ions, tribasic (e.g. orthophosphoric acid). The prefixes *per-*, *ortho-* and others will be explained under the appropriate element. Alternatively, and this is increasingly favoured by some, the *Stock Notation* may be used where possible ambiguities arise over name and structure:

HNO_3	nitric(v) acid
H_3PO_4	phosphoric(v) acid
H_2SO_4	sulphuric(vi) acid
$HClO_4$	chloric(vii) acid

The acidity of a solution depends solely upon the hydrogen ion concentration, which will be decided by the total concentration of acid dissolved and the extent to which this is ionized. Acids are broadly classified as *strong* or *weak*. This does not refer to the concentration of the acid but to the *degree (i.e. fraction) of ionization (or dissociation)* of the acid. Strong acids are virtually completely ionized in moderate concentrations (up to about 4M) while weak acids are feebly ionized.

The degree of ionization is increased by dilution of the solution. Examples will illustrate this point: hydrochloric acid is 91% ionized in decimolar (0.1 M) solution and 96.0% ionized in centimolar (0.01 M) solution, while ethanoic acid (acetic acid) is 1.3% and 4.2% ionized at these molar concentrations. Hydrochloric acid is strong and acetic acid weak. A quantitative treatment will be found in texts on Physical Chemistry.

Acidity then is characteristic of hydrogen ions. Ethanoic acid (acetic acid) will show acidic properties to a less marked degree than hydrochloric acid. The reaction between an acid and a metal will depend on the nature of the metal too. This will involve the electrode potential and overvoltage under the conditions of the experiment.

In addition, the possible insolubility of the products and the formation of complexes influence the expected course of reactions. To these must be added the effect of temperature change, always increasing the rate of reaction for a rise in temperature and influencing each of the other factors. Further complications occur when the acid has oxidizing powers, leading to secondary reactions. Magnesium reacts with very dilute nitric acid to give the nitrate and hydrogen but in more concentrated acid the gaseous product consists of oxides of nitrogen.

Most acids contain oxygen as well as hydrogen, and Lavoisier advanced a theory, based on knowledge in the year 1775, that all acids do contain it. Oxygen had just been prepared by Priestley and the name, which means *acid- former,* was given by the French chemist. When oxides of non-metals react with water, acids are formed. Sulphur trioxide yields sulphuric acid and phosphorus pentoxide, orthophosphoric acid:

$$SO_3 + H_2O \rightarrow H_2SO_4$$

$$P_4O_{10} + 6H_2O \rightarrow 4H_3PO_4$$

These oxides are called *acidic oxides* and belong to a class of compounds called *acid anhydrides.*

An *acidic oxide* reacts with bases to form salts in which the element of the oxide is in the anion,

e.g. $$Na_2O + SO_3 \rightarrow (Na^+)_2\,SO_4^{2-}$$

The Lavoisier concept was disproved about 30 years later by Davy. Hydrogen chloride had been discovered by Priestley by the action of concentrated sulphuric acid on common salt, The solution, which had typically acidic properties, was called muriatic acid (Latin, *muria*= brine). Lavoisier, working by analogy, concluded that the acidic behaviour was due to combined oxygen in muriatic acid. By heating the acid with manganese(iv) oxide (manganese dioxide), Scheele, obtained a yellow poisonous gas which he regarded as muriatic acid minus hydrogen.

In sunlight, a solution of the gas evolved oxygen leaving muriatic acid in solution, leading Berthollet to name the gas oxymuriatic acid, although the compound was not an acid. Davy had heated metals and non-metals in the gas but had not obtained oxygen or any recognized compound of oxygen and he concluded: there may be oxygen in oxymuriatic gas, but we can find none. He declared oxymuriatic acid to be an element and named it chlorine (Greek, *chloros* = greenish-yellow). Scheele's preparation may be summarized by the equation:

$$\underset{\text{manganese(iv) oxide}}{MnO_2} + 4HCl \rightarrow \underset{\text{manganese(ii) chloride}}{MnC1_2} + Cl_2 + 2H_2O$$

while Berthollet's oxygen came front the water:

$$\underset{\text{chlorine = oxymuriatic acid}}{2Cl_2} + 2H_2O \rightarrow \underset{\text{hydrochloric acid = muriatic acid}}{4HCl} + O_2 \quad \text{(net change)}$$

As we have seen, the key element in acids is the hydrogen which may be released as hydrogen ions. This is not necessarily all of the hydrogen. Only a quarter of the hydrogen in ethanoic acid, for example, is potentially acidic.

$$CH_3.COOH \rightleftharpoons \underset{\text{ethanoate ion}}{CH_3.COO^-} + H^+$$

Bases and alkalis

Acids are neutralized by bases and an *early, but still* useful, definition of the term propounded by Rouelle in referred directly to this.

A base is any substance which *will* react with an acid to produce a salt. A base may be soluble or insoluble in water, or react with water to produce a basic solution. Insoluble bases include copper(ii) oxide and magnesium hydroxide, which react with sulphuric acid:

$$\underset{\text{copper(ii) oxide}}{CuO} + \underset{\text{sulphuric acid}}{H_2SO_4} \rightarrow \underset{\text{copper(ii) sulphate}}{CUSO_4} + \underset{\text{water}}{H_2O}$$

$$Mg(OH)_2 + H_2SO_4 \rightarrow MgSO_4 + 2H_2O$$

magnesium hydroxide   *sulphuric acid*   *magnesium sulphate*   *water*

The products remain in solution.

Potassium hydroxide is a soluble base, reacting with nitric acid to form potassium nitrate solution:

$$KOH + HNO_3 \rightarrow KNO_3 + H_2O$$

Barium oxide reacts violently with water to form the soluble base, barium hydroxide, which may be used to neutralize hydrochloric acid:

$$BaO + H_2O \rightarrow Ba(OH)_2$$

$$Ba(OH)_2 + 2HCl \rightarrow BaCl_2 + 2H_2O$$

barium hydroxide   *hydrochloric acid*   *barium chloride*

A solution of barium chloride is formed in this example.

Metal oxides which neutralize acids are called basic oxides. A soluble base is commonly called an alkali. A basic oxide reacts with acids to form salts in which the element of the oxide is in the cation. Alkaline solutions, in addition to neutralizing acids, feel slimy to the touch, taste soapy and reverse the colour changes brought about by the action of acids on indicators.

Common alkalis include the hydroxides of the metals of Groups Ia and IIa of the Periodic Table (Li to Cs, Be to Ra). They are strong electrolytes because of their ionic nature and the alkaline character is due to the hydroxide, formerly hydroxyl, ion. The ionic structures dissociate in water:

$$Na^+OH^- = Na^+ + OH^-$$

$$Ba^{2+}(OH^-)_2 = Ba^{2+} + 2OH^-$$

An alkali is a compound which dissociates to produce

hydroxide ions when dissolved in water. Aqueous solutions of ammonia and its organic derivatives, such as triethylamine, are alkaline by interaction with the solvent,

$$\underset{\textit{ammonia}}{NH_3 + H_2O} \rightleftharpoons \underset{\textit{ammonium ion}}{NH_4^+ + OH^-}$$

$$\underset{\textit{triethyl amine}}{N(C_2H_5)_3} + H_2O \rightleftharpoons \underset{\textit{triethyl ammonium ion}}{N(C_2H_5)_3H^+} + OH^-$$

The presence of hydroxide joins may easily demonstrated. However, ammonia and the amines will react directly with acids without going into solution first:

$$NH_3 + \underset{\textit{hydrogen bromide}}{HB_r} \rightarrow \underset{\textit{ammonium bromide}}{NH_4^+ Br^-}$$

$$N(C_2H_5)_3 + HBr \rightarrow \underset{\textit{triethylammonium bromine}}{[N(C_2H_5)_3H^+]Br-}$$

Therefore:

A base is a compound which wiil unite with hydrogen ions.

Bronsted-Lowry Theory

Independently in J. N. Bronsted and T. M. Lowry developed a general theory of acids and bases. So far, we have regarded acids as compounds giving hydrogen ions and soluble bases as yielding hydroxide ions in aqueous solution. Such a view is simple and straightforward but with nonaqueous media, of very great importance nowadays, a more general approach is needed.

A modified concept of what is understood by the terms acid and base may be developed. An acid is any compound having a tendency to lose a proton. A base is any compound having a tendency to accept a proton. An acid and base are said to be *conjugate with one another* when related by the transfer of a proton:

$$\mathrm{Acid} \rightleftharpoons \mathrm{H^+ + Base}$$

In aqueous solution the base proton forms the oxonium ion; using hydrochloric acid:

$$\underset{\substack{\text{acid}\\ 1}}{\mathrm{HCl}} + \underset{\substack{\text{base}\\ 2}}{\mathrm{H_2O}} \rightleftharpoons \underset{\substack{\text{acid}\\ 2}}{\mathrm{H_3O^+}} + \underset{\substack{\text{base}\\ 1}}{\mathrm{Cl^-}}$$

Both hydrogen chloride and oxonium ions tend to throw off protons, so for convenience they may be labelled acid I and acid 2 respectively. The conjugate bases are the chloride ion, base 1, and the solvent, water, base 2, which has accepted protons. The equilibrium lies far to the right, so that hydrochloric acid is a strong acid.

The chloride ion must be a weak base, having little tendency to revert to the acid by accepting a proton. The anion of a feebly ionized acid clearly has a strong tendency to form the conjugate acid with protons and is therefore a strong base. Hydrocyanic acid is such an acid and the cyanide ion a strong base:

$$\underset{\substack{\text{hydrogen}\\ \text{cyanide}\\ \text{acid}\\ 1}}{\mathrm{HCN}} + \underset{\substack{\text{base}\\ 2}}{\mathrm{H_2O}} \rightleftharpoons \underset{\substack{\text{acid}\\ 2}}{\mathrm{H_3O^+}} + \underset{\substack{\text{base}\\ 1}}{\mathrm{CN^-}}$$

The strength of an acid depends on the strength of the solvent as a base—on its ability to accept protons from the acid. Solvents may be classified as proton-acceptors (protophilic), proton-donors *(protogenic)* or as neither of these (aprotic). By definition, protogenic solvents are acids and protophilic, bases. Ethanoic acid, weak in water, is strongly ionized in liquid ammonia, a base. Bases are stronger when dissolved in ethanoic acid than in a less protogenic solvent such as water. Pure water itself undergoes self-ionization in which one water molecule *protonates* another:

$$H_2O + H_2O \rightleftharpoons H_3O^+ + OH^-$$

for which the ionic product of water

$$K_w = [H_3O^+][OH^-]\ 1 \times 10^{-14} mol^2\ dm^{-6} \text{ at } 298\ K$$

$$\text{Since } [H_3O^+] = [OH^-] = 1 \times 10^{-7}\ mol\ dm^{-3}$$

Liquid ammonia has a high dielectric constant, is a poor conductor of electricity and is an ionizing

$$2NH_3 \rightleftharpoons \underset{\textit{acidic ion}}{NH_4^+} + \underset{\textit{basic ion}}{NH_2^-}$$

The proton is carried by an ammonia molecule, which becomes the ammonium ion. Ammono-acids and ammono-bases neutralize each other:

$$\underset{\substack{\textit{ethanamide}\\ \textit{(acetamide)}}}{CH_3.CO.NHH} + \underset{\textit{sodamide}}{Na^+ NH_2^-} \rightarrow$$

$$CH_3.CO.NH^-Na^+ + NH_3$$

$$\underset{\textit{ammonium chloride}}{NH_4Cl} + NaNH_2 \rightarrow 2NH_3 + \underset{\textit{sodamide}}{NaCl}$$

which corresponds to

$$HCl + NaOH \rightarrow H_2O + NaCl$$

or more accurately

$$OH_3Cl + NaOH \rightarrow 2H_2O + NaCl$$

Acid-base theory has been extended into nonprotonic systems, i.e. where protons are not encountered. Many irterhalogen compounds ionize slightly and are ionizing solvents:

$$2ICl_3 \rightleftharpoons ICl_2^+ + ICl_4^-$$

$$2BrF_3 \rightleftharpoons BrF_2^+ + BrF4^-$$

$$2IF_5 \rightleftharpoons \underset{\substack{\uparrow \\ \textit{acids}}}{IF_4^+} + \underset{\substack{\uparrow \\ \textit{bases}}}{IF_6^-}$$

From this brief survey, final definitions may be stated:

An acid is a compound which ionizes to give the cation of the parent solvent. A base is a compound which ionizes to give the anion of the parent solvent.

Normal, acid and basic salts

A salt is formed by replacement of acidic hydrogen in an acid by a metal or a basic group (such as NH_4^+). Very few of these compounds are not ionic and the term salt is sometimes restricted to those with an essentially ionic structure. For example, sodium sulphate, $(Na_+)_2SO_4$, is composed of sodium ions and sulphate ions in the ratio 2 : 1.

The sulphate ion itself is covalent internally. Ion-pair formation and various strong electrolyte effects occur in concentrated solutions so that determinations of the apparent degree of dissociation by colligative methods will indicate that ionization is apparently incomplete. Colligative properties (relative lowering of the vapour pressure, depression of the freezing-point, elevation of the boiling-point and osmotic pressure) depend on the number of particles (molecules, ions) and not on their size or chemical nature. In some are cases, e.g. CdI_2, complex ion formation occurs in solution.

The positive ion is sometimes called the basic radical and the negative ion, the acidic radical. Ammonium nitrate, $NH_4^+NO_3^-$ contains no metal, having the ammonium radical as the cation and the nitrate radical as the anion. However, such usage of the term radical is best avoided.

The term is best regarded as describing a group of atoms which occurs repeatedly in a number of different compounds. Examples are $COCl_2$, *carbonyl* chloride, $PSCl_3$, *thiophosphoryl* chloride, POCl, phos*phoryl(III)* chloride, CrO_2Cl_2, *chromyl* chloride.

When all of the acidic hydrogen has been replaced the product is called a *normal salt;* it contains neither H^+ from the acid, nor OH^- from the base:

$$2NaOH + H_2SO_4 \rightarrow Na_2SO_4 + 2H_2O$$

sodium sulphate

$$3NaOH + H_3PO_4 \rightarrow Na_3PO_4 + 3H_2O$$

trisodium orthophosphate

Both sodium sulphate and trisodium orthophosphate are normal salts. Polybasic acids may form *acid salts* in which some acidic hydrogen remains in each anion. Such salts release hydrogen ions in water being also acids. Ionization of the anions occurs:

$$HSO_4^- \rightleftharpoons H^+ + SO_4^{2-}$$

$$H_2PO_4^- \rightleftharpoons H^+ + HPO_4^{2-}$$

$$HPO_4^{2-} \rightleftharpoons H+ + PO_4^{3-}$$

Sodium hydrogensulphate is sometimes known as sodium bisulphate. Sodium hydrogencarbonate (sodium bicarbonate) and sodium hydrogensulphite (sodium bisulphite) are other examples of acid salts. As the name suggests, *basic salts* retain a portion of the base, either as oxygen or as a hydroxide (hydroxyl) group.

The normal salt may usually be formed with strong acids. The formation of basic salts is a sign of weakness in metallic nature of the participating cation of the salt. Bismuth oxide chloride, formerly oxy-chloride, BiOCI, yields the normal salt, bismuth(iii) chloride (bismuth trichloride), with hydrochloric acid,

$$BiOCl + 2HCl \rightarrow BiCl_3 + H_2O$$

and tin(ii) hydroxide chloride, [$Sn^+(OH)Cl^-$ reacts similarly

$$Sn(OH)Cl + HC1 \rightarrow SnCl_2 + H_2O$$

Basic copper(ii) ethanoate (acetate), $Cu(OH)_2$. $Cu(CH_3COO)_2$, or verdigris, white lead, basic lead(ii) carbonate, $2Pb(OH)_2$ $PbCO_3$, and malachite, basic copper(ii) carbonate, $Cu(OH)_2$. $CuCO_3$, are other illustrations.

Nomenclature for acids and salts

The prefixes and suffixes marked in Table 3 should be noted. Sodium chloride and hydrochloric acid each consist of two elements only; the suffix *-ous* denotes this for the salt while-ic is derived from the way Lavoisier named acids in connection with his oxygen theory of acids. However, note in the other cases that an acid with a name ending in -ic forms salts which have names ending in *-ate.* With one oxygen atom per molecule less than chlorous acid, HClO is called hypochlorous acid, hypo (Greek = below or lower) indicating the lower oxygen content; the salts are called hypochlorites. The prefix per (Latin = completely) is used to show a maximum of combined oxygen so that $HClO_4$ with one more oxygen atom per molecule than chloric acid, is called perchloric acid and the salts, perchlorates.

Table 3

The naming of acids and salts

Formula of acid	*Name of acid*	*Formula sodium salt*	*Name of sodium salt*
HCl	Hydrochloric	NaCl	Chloride
HClO	Hypochlorus	NaClO	Hypochlorite
$HClO_2$	Chlorous	$NaClO_2$	Chlorite
$HClO_3$	Chloric	$NaClO_3$	Chlorate
$HClO_4$	Perchloric	$NaClO_4$	Perchlorate

When an element or radical forms salts in which it exhibits more than one valency, until recently the suffix *-ous* denoted the lower valency state and -ic the higher. This now applies to certain oxo-acids only. Ferrous sulphate is now known as iron(ii) sulphate and ferric sulphate as iron(iii) sulphate where the roman numerals indicate the oxidation state.

If the *Stock Notation is* applied to acids and their anions, H_2SO_3 becomes sulphuric(iv) acid and H_2SO_4, sulphuric(vi) acid, and the respective salts sodium sulphate(iv) and sodium sulphate(vi). Alternatively, the proportion of oxygen might be indicated as in sodium trioxosulphate and sodium tetraoxosulphate, to use the salts Na_2SO_3 and Na_2SO_4 as illustrations. Note that the terms sulphuric and sulphate, used originally to specify a particular acid and its anion (salt), should now designate quite generally an acid or a negative group containing sulphur as the central atom (linked to other atoms) irrespective of its state of oxidation. While the oxidation number is stated in the first set of examples, it is implied in the second.

Table 4. shows the Stock Notation extended to those acids and their salts already named in Table 3. The alternative names should be compared.

Table 4

The stock notation for acids and salts

Formula of acid	*Oxidation number of chlorine*	*Name of acid*	*Name of anion (i.e. salt)*
HCl	-1	Hydrochloric	Chloride
HClO	+1	Chloric (I)	Chlorate (I)
$HClO_2$	+3	Chloric (III)	Chlorate (III)
$HClO_3$	+5	Chloric (V)	Chlorate (V)
$HClO_4$	+7	Chloric (VII)	Chlorate (VII)

While the terms hydrochloric and chloride remain unchanged, chloric and chlorate have become general family terms, requiring the addition of oxidation numbers to be specific. The use of the prefix *-ortho, as* in orthophosphoric acid, H_3PO_4, will have been noticed. *Ortho (Greek* = correct) is used to indicate the presence of the maximum quantity of the elements of water uniting.

Loss of water from each molecule yields metaphosphoric acid, HPO_3, where meta (Greek = after, Le. H3PO4) shows the relationship. An intermediate acid formed by the action of heat on orthophosphoric acid is pyrophosphoric acid, $H_4P_2O_7$, where pyros (Greek fire) indicates the mode of formation.

Salt hydrolysis

A normal salt is the product of the neutralization of an acid by a base, all of the acidic hydrogen of the acid being replaced. If the salt is soluble, tests with indicators may show that the pH (= $-\log_{10}[H^+]/\text{mol dm}^{-3}$) is not 7 at 25°C; neutralization has not produced a neutral solution! The pH of a salt solution depends on the relative strengths of the combining acids and bases. By interaction with the solvent, the equilibrium,

$$2H_2O \rightleftharpoons H_3O^+ + OH^-$$

is disturbed, with the result that the concentrations of hydrogen (oxonium) ion and hydroxide ion are no longer equal and the solution no longer neutral. This phenomenon is called salt hydrolysis and will be examined under two headings.

Hydrolysis of salts of soluble bases

Salts of strong acids with strong bases are not hydrolysed and neutral solutions are formed. However, hydrolysis occurs where weak acids and weak bases are used. Salts of weak acids and strong bases, e.g. potassium cyanide solution:

$$(K^+ +)\ CN^- + H_2O \rightleftharpoons HCN + OH^-\ (+ K^+)$$

Hydrocyanic acid is feebly ionized. The solution contains an excess of hydroxide ions over hydrogen ions. It is alkaline to litmus (pH > 7.).

Salts of strong acids and weak bases, e.g. ammonium chloride solution:

$$(Cl^- +)\ NH_4^+ + H_2O \rightleftharpoons NH_3 + H_3O^+\ (+Cl^-)$$

Aqueous ammonia is a weak base with only a small tendency to compete for the proton. There is an excess of oxonium (hydrogen) ions and the solution is acidic to litmus (pH < 7).

Salts of weak acids and weak bases, e.g. ammonium ethanoate (acetate) solution:

$$(NH_4^+ +)CH_3COO^- + 2H_2O$$

$$NH_3 + H_3O^+ + CH_3COOH + OH^-$$

which becomes

$$NH_4^+ + CH_3COO^- \rightleftharpoons NH_3 + CH_3COOH$$

Although each ion is involved in hydrolysis, the final acidity of the resulting solution will depend on the relative extent of the two processes. Actually, in this case the processes of salt hydrolysis cancel each other out and the solution is neutral.

Hydrolysis of salts of insoluble bases

Metal hydroxides formed by elements other than those of the metals of Group I (alkali metals) and II (alkaline earths) of the Periodic Classification are weak bases, being virtually insoluble in water. Salts of these bases are often acidic in solution. Copper(ii) sulphate, aluminium sulphate and iron(iii) chloride are examples. Consider aluminium sulphate solution.

The aluminium ions are hydrated as $Al(OH_2)_6^{3+}$ the

hexaquoaluminium ion, in which the water molecules are probably linked by co-ionic bonds from oxygen to metal in octahedral formation (i.e. producing an octahedral shape to the cation) although the bonding might be electrostatic in nature. Competition between hydrated aluminium ions and solvent molecules leads to the loss of protons:

$$Al(OH_2)_6^{3+} + H_2O \rightleftharpoons Al(OH_2)_5(OH)^{2+} + H_3O^+$$

$$Al(OH_2)_5(OH)^{2+} + H_2O \rightleftharpoons Al(OH_2)_4(OH)_2^+ + H_3O^+$$

$$Al(OH_2)_4(OH)_2^+ + H_2O \rightleftharpoons Al(OH_2)_3(OH)_3 + H_3O+$$

hydrated aluminium oxide or hydroxide

Unless alkali is added, the hydrolysis will not go as far as the third stage. Iron(iii) chloride solution may become reddish-brown on storage indicating the presence of hydrated iron(III) oxide in colloidal solution. The addition of acid reduces hydrolysis and gives a solution with a yellow tinge. The products of hydrolysis of the hexaquoiron(iii) ion, $Fe(OH_2)_6^{3+}$ will be

$$Fe(OH_2)_5(OH)^{2+}, Fe(OH_2)_4(OH)_2^+$$

and $Fe_2O_3 . xH_2O$ or hydrated $Fe(OH)_3$

This type of hydrolysis is pronounced with hydrated ions of high charge (Al^{3+}, Sb^{3+}, Bi^{3+}, Fe^{3+}), much less so with bivalent ions (Zn^{2+}, CU^{2+} are hydrolysed while Fe^{2+} is not) and not at all with ions of single charge. The key to the discussion lies in the polarizing the power (Fajans' Rules) in determining the acidity of $[M(H_2O)_6]^{n+}$ ions from the Lowry-Bronsted view point. This polarizing power may be expressed as ions of single charge (e.g. Ag^+). charge/radius, the strong bond formed between the metal atom (aluminium) and oxygen weakening the O—H bond of the attached water molecules leading to dissociation and an acidic solution.

Chapter 7

Mechanism and Reactivity

Many reactions in inorganic chemistry take place very rapidly to give the products with the greatest thermodynamic stability. This contrasts with organic chemistry where reactions are generally quite slow, and give the kinetically stable product rather than the most thermodynamically stable one. For this reason, mechanistic inorganic chemistry developed more slowly than mechanistic organic chemistry, but increasing interest in forming products other than the thermodynamically most stable, in speeding up or catalysing slow reactions, and the development of new methods of studying fast reactions has given a considerable boost to the subject. All chemical reactions will have a limited rate, if only because of the necessity of approach of two molecules to produce the reaction (even in collisionally excited unimolecular reactions). This diffusion rate is extremely fast in liquid and gas phases, but much slower (though non-zero) in the solid state.

In gas and liquid states reactions are normally slower than the diffusion rate, as a result of the activation energy of the reaction.The traditional view of a chemical reaction shown in Fig. 1a involves passage through an excited transition state, where nuclear positions in the reactants have been distorted in such a way as to favour the direct passage to the products. The lower the activation energy the more accessible the *transition state*, and the faster the reaction. Although the transition state does not actually exist (it is at a maximum of potential energy), it is useful to consider the structure of this hypothetical entity since we can then estimate how energetically accessible it will be. Figure 1b shows a two step reaction with the formation of an intermediate, of high energy, between two transitions. Although it may have high energy, the intermediate is at the minimum of a potential well, and may be experimentally detectable during the reaction.

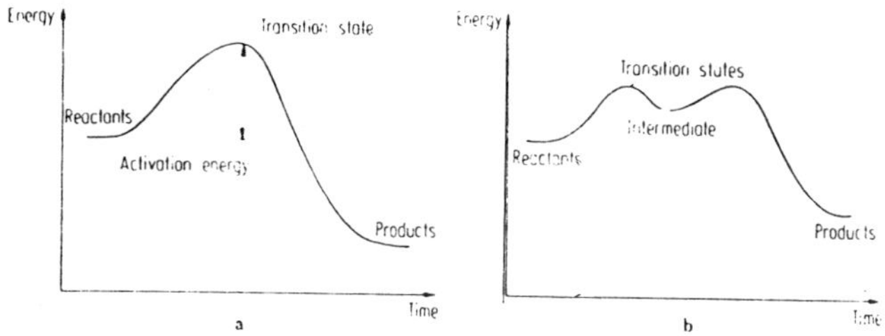

Fig. 1: The variation of energy of a reactant mixture viewed along, the reaction coordinate. (a) A simple reaction. (b) a reaction with formation of an intermediate

The formation of the transition state involves extension or compression of various bonds in the reacting molecules, and maybe the formation or rupture of chemical bonds. There are two useful ways of looking at this—we may either concentrate on one or two localised bonds which will be broken or formed, or we may regard all the various nuclear motions (the stretching or compressing of interatomic distances) as a vibrational mode of the transition state, just as we would discuss the nuclear motions of an ordinary molecule.

The first method is generally simpler but the second is more general. In both cases we are seeking to estimate the energetic barrier to the formation of the transition state, by consideration of the molecular structures of the reactants,

Entropy of activation

While the activation energy is generally the dominating factor in a chemical reaction it is not invariably so. Transition state theory gives an expression for the rate constant

$$k = K\frac{RTe}{Nh}e^{-E_a/RT}e^{\Delta S\neq/R} \tag{1}$$

where k = transmission coefficient (generally put equal to 1 since its calculation is very complicated)

E_a = Arrhenius activation energy

$\Delta S^{\neq}$ = entropy of activation.

The enthalpy of activation in solution is given by $\Delta H^{\neq} = E_a - RT$. The entropy of activation is regarded as the difference in entropy between reactants and the transition state. One might expect that a transition state involving association between two reactant particles would have a more negative entropy of activation (as a result of loss of translational entropy) than a dissociative mechanism; there is some evidence to support this, but solvation effects can dominate $\Delta S^{\neq}$, especially when the transition state is more polarised than the reactants (as a result, perhaps, of a partial ionic dissociation).

The solvent molecules are then more strongly attracted to the complex with consequent loss of entropy. Although we shall concentrate on activation energies, it should be remembered that differences in activation entropies are sometimes important. Thus $[Au(dien)Cl]^{2+}$, reacts with methanol 3×10^4 times faster than the isoelectronic, isostructural $Pt(dien)Cl^+$, but the $\Delta H^{\neq}$ for the reactions are very similar, the rate differences coming from $\Delta S^{\neq}$ differences. We will initially discuss this topic on a localised bond model, and return to an M.O. approach later. If we accept that a localised bond involves one electron pair, then the fission of this bond can occur in two ways:

— Each atom involved in the bond holds onto one electron, producing two radicals, each with an unpaired electron (*Homolytic fission*);

— The bond electron pair stays localised on one of the atoms of the bond (*Heterolytic fission*). This need not necessarily involve the formation of ions—the heterolytic fission of HCl produces H^+ and Cl^-, but the fission of the B—N bond in H_3NBF_3 gives NH_3 and BF_3 with the bonding pair of the B—N bond now a lone pair on NH_3. Heterolytic fission is clearly much more probable for polar chemical bonds.

The corresponding processes of bond formation are quite straightforward—radicals may either dimerise (quite rare as their concentration is generally low), or add to a non-radical molecule to give a new radical which can either dimerise, or eliminate some other radical species, leaving behind a new molecule. The corresponding equations are

$$2A\cdot \rightarrow AA$$

$$A\cdot + B \rightarrow AB^\cdot \rightarrow C + D\cdot$$

The free radical is a rather old concept in mechanistic chemistry and, historically, has always an implied reactivity. It should however be remembered that a species with one or more unpaired electrons, although a radical, is not necessarily very reactive: Mn^{2+} generally has five unpaired spins but shows no tendency to undergo radical mechanism reactions.

Heterolytic bond formation implies the donation of an electron pair from one atom to form a chemical bond with another—that is, a Lewis acid/Lewis base reaction.

Nucleophiles and electrophiles

It is customary, when discussing heterolytic bond formation, to classify reagents as *nucleophiles*, tending to react with positive centres and acting as electron donors, or *electrophiles*, tending to act as electron acceptors and reacting with negative centres. This is an extremely useful classification for discussing reactivity; as a general rule, species which are Lewis bases are nucleophiles, and those which are Lewis acids are electrophiles.

Nucleophilicity and electrophilicity are however measured from rates of reaction, and scales of nucleophilicity are quite different from scales of basicity—this was in fact one of the starting points for hard and soft acid-base theory. Thus iodide is a weak base but quite a good nucleophile. As one might expect, nucleophilicity and electrophilicity are extremely solvent dependent. One can, however, often make very useful predictions of reaction products simply by estimating the charge distributions in two reactant molecules and interacting the most nucleophilic atom of one reactant with the most electrophilic atom of the other.

To take a trivial example, the hydrolysis of $(CH_3)_2SiCl_2$ may reasonably be expected to involve attack by nucleophilic water on the electropositive silicon atom, whose electrophilicity will be increased by the two electronegative chlorine atoms. This sort of reasoning is of course the basis of a vast amount of mechanistic organic chemistry; it is somewhat less satisfactory in inorganic chemistry where it is not always simple to identify which centres are nucleophilic or electrophilic.

Molecular orbital theories of chemical reaction

The heterolytic/homolytic approach to bond formation or

breaking has several disadvantages—quite apart from the fact that it regards chemical bonds as localised, it is often difficult to relate the mechanism to the finer details of electronic structure that we have elaborated in previous chapters. It also gives us little exact understanding of how a particular bond is broken, especially for the heterolytic fission of covalent or fairly covalent bonds.

Finally, it is rather hard to explain reactions where several bonds are made and broken at the same time—for example, the addition of hydrogen to a molecule may involve the simultaneous formation of two sigma bonds—such *concerted* reactions are particularly common in organic chemistry, but are by no means unknown in inorganic chemistry. The nucleophilic Cl^- approaches the HCl molecule and overlaps with the acceptor orbital, the empty σ^* orbital mainly localised on the hydrogen atom. This donation establishes a new HCl bond and weakens the first due to occupation of the σ^* antibonding orbital, until the original, covalently bound Cl is lost as Cl^-. The approach of the Cl^- ion thus forms a new HCl bond at the same time as the old bond is weakened.

Symmetry in chemical reactions

It will be useful to discuss exactly what we mean by symmetry as applied to reacting species. The initial symmetry of the reactants (or the final symmetry of the products) will be of little interest since it is usually lowered during a chemical reaction. It is however frequently found that some elements of symmetry are maintained during a chemical reaction.

In the reaction $Cl^- + HCl$, cylindrical symmetry was maintained, and in the reaction $F_2 + Br_2 \rightarrow F_2BrBr$ we may see that a two-fold axis (along the Br—Br bond) and two perpendicular planes of symmetry parallel to this axis are maintained. It is the symmetry elements maintained throughout the reaction that are important since the orbitals involved cannot change their symmetry with respect to these operations. We may use this conserved symmetry to classify the orbitals and predict their mixing. It is quite easy to show that any vibration responsible for the reaction or mixing must be totally symmetric (i.e. A_1) with respect to the symmetry maintained throughout the reaction.

Thus, for the Cl^- + HCl reaction, the symmetry maintained is $C_{\infty v}$; the vibration corresponding to the nuclear motions in the reaction is the asymmetric stretch, which preserves cylindrical symmetry and is totally symmetric with respect to it (type σ). We may give a more complicated example from organic chemistry—the ring opening of cyclobutene. Cyclobutene and butadiene both have C_{2v} symmetry. In cyclobutene we wish to break one carbon-carbon σ bond and one π bond, to produce two π bonding M.O. in butadiene. The symmetries of the relevant orbitals are shown in Table 1.

In the C_{2v} point group we notice that the occupied orbitals of product and reactant do not correlate and that there is no means of weakening chemical bonds in cyclobutene by mixing an occupied bonding M.O. with an empty bonding M.O. (e.g. σ^* with π). We conclude that C_{2v} symmetry cannot be maintained throughout the reaction, This might also have been predicted by noticing that the hydrogen atoms of the methylene groups must be moved from out of the carbon plane (in cyclobutene) into the carbon plane (in butadiene) and that this is impossible if full symmetry is maintained. It is possible to conserve either the two fold axis of symmetry (a conrotatory reaction) or the xz plane of symmetry (a disrotatory reaction).

The point groups corresponding to these paths are C_2 and C_s respectively. Inspection of the C_{2v} character table enables us to classify the M.O. as symmetric (type A or A') or antisymmetric (B or A") with respect to these symmetries. We see that if C_2 symmetry is maintained, σ and π^*, and σ^* and π can mix (and thus weaken the desired bonds) and that the occupied orbitals in reactant and product correlate (i.e. have the same symmetry A + B). The reader should be able to see that this is not the case for the C_s point group. We thus predict a conrotatory reaction conserving C_2 symmetry as more probable. Isotopic labelling of the methylene protons confirms this prediction.

Table 1 Relevant M.O.s of cyclobutene butadiene

M.O.	C_{2v}	C_2	C_s
cyclobutene			
σ^*	B_2	B	A"
π^*	A_2	A	A"
π	B_1	B	A'
σ	A_1	A	A'
butadiene			
π	$A_2 + B_1$	A + BA' + A"	

What exactly is the mechanism of lowering of symmetry that enables the reaction to take place? Since there is only one molecule involved in the reaction, it will not be the approach of a second molecule (leading to overlap) but must be a distortion of the nuclear position of cyclobutene—that is, a molecular vibration. We may predict the symmetry of this vibration since we know that it maintains the C_2 axis and must therefore be symmetric with respect to it, but that it destroys the two mirror planes and is thus antisymmetric with respect to these. It must therefore be an A_2 vibration.

We might also have predicted this by second order Jahn-Teller effect arguments: if we wish to mix σ and π^* (or π and σ^*) of cyclobutene, we must find a vibration of symmetry Γ such that the direct products $\Gamma_\sigma \oplus \Gamma \oplus \Gamma_{\pi^*}$ and $\Gamma_\pi \oplus \Gamma \oplus \Gamma_{\sigma^*}$ are totally symmetric—inspection shows that the A_2 vibration will accomplish this (for the first case we have $A_1 \oplus A_2 \oplus A_2 = A_2 \oplus A_2 = A_1$). Knowing the vibrational mode which promotes the reaction, we may then find the elements of symmetry which are preserved and predict the reaction product. This sort of argument may be used to explain the fluxionality of five coordinate species such as PF_5 and $Fe(CO)_5$.

In these compounds the highest occupied orbitals have e' symmetry and there is an empty a_1 orbital at slightly higher energy. The vibrational mode producing the scambling has E' symmetry and since $E' \oplus E' \oplus A_1$ has a totally symmetric component, the vibration will allow a mixing of the e' and a_1' orbitals which will lower the energy of the distortion and thus encourage the scrambling.

Orbital correlation

Symmetry considerations will tell us if it is possible for different orbitals to mix, but perturbation theory told us earlier that the extent of mixing will be determined by the difference in energy between the orbitals. We must therefore consider not only the symmetry but also the orbital energies of the reactants. If we wish two orbitals to be mixed during the course of a chemical reaction, their energies should be fairly close this is the basis of Fukui's frontier orbital theory, frequently used in organic chemistry, which postulates that the majority of reactions take place at the position of, and in the direction of maximum overlap of the *highest occupied* M.O. (the HOMO) and the *lowest unoccupied* M.O. (the LUMO) of the reactant molecules.

This clearly combines overlap and energetic considerations as required by perturbation theory, but its application in inorganic chemistry is limited by the fact that the frontier orbitals (i.e. HOMO and LUMO) are not always obvious, several close-lying M.O. being common. In some cases it is known that M.O. close to, but not actually the HOMO or LUMO are of dominating importance.

Let us summarise the points made above:

1) Bond formation or breaking arises from the mixing of atomic and/or molecular orbitals at reaction centres. This mixing may result from overlap of orbitals of approaching particles, or from mixing induced by vibrational distortion (the second order Jahn-Teller effect). We may use the symmetry conserved during the course of the reaction to predict the possibility or impossibility of mixing.

2) The mixing of orbitals will always be greatest when their energy is closest.

3) The occupied orbitals of the reactants and products should correlate. This condition requires that the electrons displaced during the reaction should move smoothly into the lowest energy orbitals of the product.

Mechanisms where mixing is difficult or where orbitals do not correlate will have higher activation energies, and are sometimes said to be forbidden. We have emphasised the importance of mixing of orbitals in bond formation and rupture, but this needs two qualifications: firstly, the mixing criterion will clearly be most important when the bonds depend strongly on overlap, that is when they are quite covalent; secondly, the change of orbitals and energy due to mixing corresponds to a second order perturbation, and we have ignored any first order term.

Neutral, non-polar molecules will not interact strongly unless some mixing takes place, but for reactions between ionic or polar species, the electrostatic potentials of charged fragments may appreciably change the energy of the system without orbital mixing; this corresponds to the first order perturbation term. Thus the approach of Cl^- to $H^{\delta+}$—$Cl^{\delta-}$ will involve Cl^- . . . . . $H^{\delta+}$ attraction and Cl^- . . . .$Cl^{\delta-}$ repulsion without any overlap.

Discussion of reactions of the many highly ionic species in inorganic chemistry in terms of overlap is somewhat pointless. However, most reactions of ionic species take place in solution where the ions are dissociated, and strongly solvated. Overlap and mixing approaches are particularly useful for non-ionic species such as organo-metallic compounds. Generally speaking, the predictions of allowed or forbidden reactions are followed in the sense that allowed reactions will have lower energies of activation than similar reactions which are forbidden. It is not possible to compare very different types of reaction since allowed reactions do not necessarily have low absolute heats of activation, nor do all forbidden reactions have high or experimentally inaccessible activation energies.

Predictions are more qualitative than quantitative, and they frequently permit the explanation of the stereochemistry of the product. Small molecules containing light elements are often the most successful field of application of these theories, as is the study of the possible catalysis of reactions of such species. Other selection rules for chemical reactions may be added to those discussed above. One of the more obvious is that the total spin should not change during the reaction ($\Delta S = 0$).

The necessity of changing spin will slow down a chemical reaction, but as with absorption spectra, a little bit of spin-orbit coupling can get round this problem. Other rules are more empirical: the useful principle of least motion states than the lowest activation energy for a reaction is the one that requires the least motion of the nuclei and the least disruption of the original electronic distribution. This is intuitively reasonable but suffers from the difficulty of measuring the motions involved—nonetheless it is useful as a rule of thumb. Pearson suggests rewriting it to demand the retention of the maximum number of symmetry elements.

The principle of least motion gives a useful starting point for the suggestion of a mechanism which may then be tested for orbital symmetry matching etc. Another useful rule which is borne out by confirmed mechanisms is that multiple bonds are broken one at a time and thus that two distinct steps will be needed to break a double bond. It is clear that the picture of chemical reactions outlined in this section is very convenient for calculation of most favoured reaction paths and many such calculations have been made sometimes giving good agreement with experimental results.

On the qualitative level, it has proved possible to give rationalisations of mechanisms which are otherwise difficult to explain, starting from fairly simple assumptions of the molecular structure.

Analysis of reaction mechanisms

We may break up almost any complex chemical reaction into a series of simple steps which may frequently be distinguished experimentally. We may then relate the simple steps to the molecular structure as outlined in the previous section. It is extremely useful to have a system of classification of elementary steps so that we can systematically investigate the susceptibility of a given compound to various different reactions.

The system given here is close to that given by Tobe in his book on reaction mechanisms. An atom in a molecule will be susceptible to two forms of chemical change: a change in the coordination shell or a change in the oxidation state. Tobe also considers a third possibility, of reactions between the ligands bonded to the atom—this is very useful when discussing the coordination chemistry of metals, but it is clear that this third group may involve reactions where the central atom undergoes no chemical change, the reactive centres being elsewhere. Many reactions can be classed under two or more headings—this is a slight inconvenience, resulting from the fact that our classification is a human artefact, not a neutral division. We will adopt the following classifications:

A. *Change in coordination shell*

1. Addition—increase in coordinate number
2. Elimination—reduction in coordination number
3. Substitution—exchange of a ligand with a free ligating species
4. Change in geometry—a change in ligand-atom-ligand bond angles without actual breaking of a ligand-atom bond—for example a square planar-tetrahedral conversion
5. Change in ligand arrangement—for example a cis-trans isomerisation
6. Reaction between ligands (e.g. addition) involving a change in coordination shell.

B. *Electron-transfer reactions*

1. A simple loss or gain of electrons involving no change in coordination shell
2. A redox reaction involving a change in coordination shell

We may now discuss these elementary reactions separately, We shall label M the atom being oxidised or reduced, or whose coordination shell is being altered, the ligands involved being denoted L.

Addition $ML_n + L \rightarrow ML_{n+1}$

The acceptor or donor orbital may be a molecular orbital. In interactions of an electrophile with a π orbitals and of a nucleophile with a π^* orbital are shown—the structures of the products are different in the two cases as a result of the different spatial distributions of the M.O. involved. Radicals may act similarly to nucleophiles *or* electrophiles—if the partially filled orbital has high energy the radical will tend towards nucleophilic behaviour, if the orbital is strongly bound (as in F.) it will act as an electrophile.

Elimination $ML_n \rightarrow ML_{n-1} + L$

Elimination may simply be regarded as the reverse of the addition reaction—it will be favoured by a crowded, distorted coordination shell. Heterolytic elimination will be favoured by polar solvents stabilising the Lewis acid and Lewis base formed. Homolytic elimination (i.e. dissociation) from a non-radical species is difficult and generally requires activation (either thermal or photochemical)—elimination of a radical from a larger radical is much easier. Radical elimination is also sometimes found after an electron transfer reaction as in the case of the reaction between $[Cr(OH_2)_6]^{2+}$ + $C_6H_5CH_2Cl$ giving $(Cr(OH_2)_5Cl]^{2+}$ and $C_6H_5CH_2$. The $C_6H_5CH_2Cl$ is reduced by the chromium (II) giving, it is believed, a chloro complex of Cr(III) and the relatively stable benzyl radical.

Substitution reactions $ML_n + L' \rightarrow ML_{n-1}L' + L$

Substitution reactions are perhaps the most studied field of mechanistic inorganic chemistry; particularly popular have been nucleophilic substitution reactions at metal centres. Since most of these reactions in inorganic chemistry are fast, interest was initially concentrated on those complexes which react slowly enough to be followed by classical means—this accounts for the apparent obsession of such studies with complexes of chromium (III) and cobalt (III) (both octahedral) and square planar Pt(II) and Pd(II). The increasing availability of fast kinetic methods is expanding the field of study. In any, substitution reaction we need to break one bond and form another—the different mechanisms involved differ in the relative timing of these two steps.

The associative (A) mechanism involves an initial addition of the substituting ligand L' followed by the loss of the replaced ligand L—in the limiting case an intermediate of increased coordination may be isolated. The dissociative (D) mechanism involves initial elimination of L followed by addition of L'; here it may be possible to find an intermediate of reduced coordination number.

Finally, we may imagine an interchange mechanism where the bond formation and bond breaking steps are almost simultaneous and affect each other—this may conveniently be divided into two classes—the interchange (associative) or I_a mechanism where approach by the incoming ligand slightly precedes the weaking of the bond to the outgoing ligand, and the interchange (dissociative) or I_d mechanism, where slight bond weakening precedes substituent approach. These mechanisms may also be classified using the Hughes Ingold system popular in organic chemistry, giving the following equivalencies:

A	$S_\alpha 2$(limiting)
I_a	$S_\alpha 2$
I_d	$S_\alpha 1$
D	$S_\alpha 1$(limiting)

where $\alpha = N$ for a nucleophilic substitution
E for a electrophilic substitution
H for a homolytic (radical) substitution.

Nucleophilic substitution is extremely common, and all of the mechanisms mentioned above are known. With the possible exception of certain low valence states (where the bonding is almost certainly highly covalent) all metal atoms in compounds are electrophilic centres, and ligand substitution involves the exchange of nucleophiles. Electrophilic substitution is less common, although proton exchange and attack by metal ions on negatively charged centres may be regarded as electrophilic attacks. It is often possible to regard nucleophilic substitution on one atom as electrophilic substitution on another—tradition favours the nucleophilic point of view. For example the reaction between $[Co(III)(NH_3)_5(OH)^{2+}$ and N_2O_3 is thought to proceed as shown below

$$\begin{array}{c} \quad\quad ON(NO_2) \\ \quad\quad | \\ [H_3N)_5\ Co—O—H]^{2+} \end{array} \rightarrow [H_3N)_5\ CoONO]^{2+} + H+ NO_2^-$$

which may be regarded as nucleophilic substitution of nitrogen or electrophilic substitution of oxygen. It is also worth noting that there is no actual substitution of the cobalt, despite the apparent HO^-/NO_2^- exchange. Radical substitution is also well known, especially for the p block elements. Given our earlier remarks about radical elimination, it is unlikely that a D mechanism will be adopted, but an associative S_H2 mechanism has been found for radical attack on organo boron compounds and inversion has been found for attack on phosphorus compounds.

If we restrict our discussion to nucleophilic substitution, there are some regular features in the mechanisms adopted. For main group elements, tetrahedral species shown all four mechanisms, but the smaller atoms favour the more dissociative reactions (carbon shows I_a I_d and D behaviour) while the heavier elements, which often show Lewis acidity, show a slight partiality for associative mechanisms.

Transition metals in high oxidation states (e.g., MnO_4^-) show a tendency to associative reactions (dissociative reactions might involve the breaking of a double bond in one step), but the tetrahedral d^{10} complexes of metals in low oxidation states give dissociative reactions. Square planar 16 electron complexes give additive reactions. Five coordinate species have not been greatly studied but, with the possible exception of 18 electron transition metal species which might be expected to be dissociative, no definite preferences have been established.

Octahedral complexes generally give dissociative reactions, although there is some reason to expect associative reactions in high oxidation states, especially where species with C.N. greater than 6 may be isolated. There is little data on species with high coordination number (lanthanide or alkali metal ions in solution).

Change in geometry without bond breaking.

In the case of the nickel (II) tetrahedral-square planar inversion the reaction is forbidden since it involves a spin change of 2 units, but this is presumably overcome by spin-orbit coupling interactions. However, the relatively slow rate (10^6 sec^{-1}) by the standards of fluxional molecules, and the activation energy (45 kJ/mole) suggest that the reaction is at least being slowed down. For three coordinate pyramidal species, the inversion vibration produces the other optical isomer and thus leads to racemisation. For nitrogen this racemisation is rapid, but phosphines are slower, and salts of the pyramidal ions R R' R" S^+ were resolved using classical methods by Pope, Tetrahedral carbon or silicon cannot racemise this way; the most probable route would be through a square planar intermediate, but an orbital correlation diagram shows this to be unfavourable, and that there are no low lying empty orbitals which may be mixed in by the vibration.

Change in ligand arrangement with bond rupture.

In its simplest form this is an elimination reaction followed by an addition reaction leading to a new product. This may be a more favourable route to isomerisation than the distortion of the initial molecule without bond rupture.

Thus, in the racemisation of $[Co(en)_3]^{3+}$ one Co—N bond will break, followed by rearrangement of the 5-coordinate species, and the recombination will give either enantiomer, leading to a racemisation. In a similar way *cis*-isomers may be partially converted to *trans*-isomers and so on. For the particular case of the racemisation of metal complexes such as $[Co(en)_3]^{3+}$ there is often a choice of a bond rupture or a intramolecular twist mechanism (class A4) passing through a trigonal prismatic intermediate.

Both routes have beef) found experimentally—dithiocarbamate ($R_2NCS_2^-$) chelates appear to be particularly attached to the intramolecular mechanism. This agrees well with the observation of the formation of trigonal prismatic complexes with certain metals by this ligand.

Reactions between ligands involving a change in the coordination shell

The proximity of two ligands coordinated to a central atom may encourage them to react with each other, and this may involve a change in coordination of the central atom: a simple example is the ligand migration reaction

$$\begin{array}{c} L \\ | \\ M—L' \rightarrow M—L'—L \end{array}$$

The central atom M may in fact encourage this coupling reaction, acting as a catalyst for an otherwise difficult reaction. Another example is the reductive elimination reaction where the M—L' bond is also broken to give a free L—L' molecule. Both these reactions are discussed later in the chapter.

Electron transfer reactions

Many redox reactions are simple electron transfer reactions, but as a result of the difficulties associated with the definition of oxidation state, some redox reactions involve no electron transfer. Thus the reaction $SO_3^{2-} + OCl^- \rightarrow SO_4^{2-} + Cl^-$ might be regarded as a nucleophilic substitution (at oxygen) of SO_3^{2-} for Cl^-; we would also class this as a redox reaction, yet it is clearly different from the reaction

$$\underset{3d^5}{Fe^{3+}(aq)} + \underset{3d^4}{Cr^{2+}(aq)} \rightarrow \underset{3d^6}{Fe^{2+}(aq)} + \underset{3d^3}{Cr^{3+}(aq)}$$

where we may definitely identify an electron transfer, Similarly, oxidative addition and reductive elimination, while they involve a redistribution of electron density, only involve electron transfer if a highly ionic model of the bonding is used. In this section we will discuss those reactions where the transfer of one or more electrons is the dominant feature. Such reactions generally involve the transfer of electrons from or into weakly bonding, antibonding or non-bonding orbitals such as transition metal d orbitals or the non-bonding s orbitals of certain main group elements. Bond formation or rupture may occur in these reactions but is of secondary importance. We will discuss first those reactions where bonds are not made or broken before, during or after transfer.

Electron transfer involving no change in coordination sphere
This reaction consists of a simple electron jump between two species with no formation of a bond between them. An example is the oxidation of $(Fe(CN)_6]^{4-}$ by $[IrCl_6]^{2-}$:

$$[Fe(CN)_6]^{4-} + [IrCl_6]^{2-} \rightarrow [Fe(CN)_6]^{3-} + [IrCl_6]^{3-}$$

This type of reaction is referred to as an outer sphere electron transfer as the reductant remains in the outer coordination sphere of the oxidant (and vice versa). The electron is assumed to 'tunnel' through a potential barrier between the configuration before the reaction, and that after the reaction. Although the electron transferred is only weakly involved in chemical bonding its departure (or arrival) will have an effect on the energy of the species. Thus the removal of an electron from the t_{2g} shelf of $(Fe(CN)_6]^{4-}$ would normally lead to a shortening of the Fe—CN bond (as for WF_6^-).

If the electron jump is assumed to obey the Franck-Condon principle, then the $[Fe(CN)_6]^{3-}$ produced will be in a vibrationally excited state. Similar reasoning shows that $[IrCl_6]^{3-}$ will also be formed in a vibrationally excited state, the Ir—Cl bond length of $[IrCl_6]^{2-}$ being too short for $[IrCl_6]^{3-}$. This additional vibrational energy will provide an energy barrier to the reaction, which can be partially overcome by distorting (or 'rearranging') the two complexes before the electron transfer. The principle of microscopic reversibility which requires that the reverse reaction involves the same distortions of internuclear distances as the forward reaction leads us to conclude that the oxidant and reductant are distorted to a configuration intermediate between their structures before and after the reaction. This distortion requires an activation energy referred to as a *Franck-Condon barrier*.

Various models for calculating activation energies for outer sphere reactions have been established (notably by Marcus) allowing for ligand rearrangement, repulsion or attraction between the two reacting species, changes in solvation, and the total free energy change of the reaction. Their predictive power is fairly satisfactory, and they confirm the general view of this reaction. The initial approach of the oxidant and reductant is generally fairly rapid and electrostatic repulsion effects (e.g. between $[IrCl_6]^{2-}$ and $[Fe(CN)_6]^{4-}$) not very important, although catalysis may be important here.

The rearrangement, or Franck-Condon, barrier is often quite considerable. This is particularly true when the orbital filled or emptied has an appreciable bonding or antibonding role, since the change in its occupancy will produce an appreciable change in equilibrium bond lengths, and thus an appreciable Franck-Condon barrier.

Thus removal of an electron from the e_g^* d levels of an octahedral complex will require a larger rearrangement energy than removal of an electron from the non-bonding t_{2g} orbitals, even though the ionisation energy of the t_{2g} shell is, in principle, higher. Large negative free energy changes for the reaction may help to reduce the importance of the Franck-Condon barrier. Finally, for the tunnelling to take place, it appears to be important that there should be some slight overlap between the donor and acceptor orbital although the actual extent of the overlap may be rather small.

In the case of $[IrCl_6]^{2-}$ it will be recalled that the acceptor orbital of the t_{2g} set is partially delocalised over the chlorine ligands (from ESR results) and will thus be able to overlap weakly with reducing agents. Outer sphere electron transfer reactions vary enormously in speed, from only slightly slower than diffusion control ($10^9 s^{-1}M^{-1}$) to rather slow ($10^{-4} s^{-1}M^{-1}$). They are particularly slow for reactions where a change in spin state and electronic configuration is involved, as for example, for the Co(III) low spin—Co(II) high spin reaction.

Electron transfer involving a change in coordination shell

There is now abundant evidence for an electron transfer mechanism involving the initial formation of a complex between the oxidant and reductant in which at least one ligand is bound to both initial species. This mechanism, since it involves a merger of two coordination shells, is referred to as an inner sphere mechanism. The classical example, due to Taube, is the reduction of $[Co(NH_3)_5Cl]^{2+}$ by aqueous Cr^{2+} for which the mechanism is assumed to be

$$[(H_3N)_5CoCl]^{2+} + [Cr(OH_2)_6]^{2+} \rightarrow$$
$$[(H_3N)_5Co^{III}\text{-}Cl\text{-}Cr^{II}(OH_2)_5]^{4+} + H_2O \qquad (1)$$

$$[(H_3N)Co^{III}\text{-}Cl\text{-}Cr^{II}(OH_2)_5]^{4+} \rightarrow$$
$$[(H_3N)_5Co^{II}\text{-}Cl\text{-}Cr^{III}(OH_2)_5]^{4+} \qquad (2)$$

$$[(H_3N)_5Co^{II}\text{-}Cl\text{-}Cr^{III}(OH_2)_5]^{4+} \rightarrow$$
$$Co^{2+}\,aq + 5NH_3 + [Cl\text{-}Cr^{III}(OH_2)_5]^{2+} \qquad (3)$$

Steps 1 and 3 are simple substitution steps (which may influence the rate of reaction as a whole), and step 2 is the actual electron transfer mediated by the bridging ligand (Cl). The speeds of steps 1 and 3 are clearly dependent on the susceptibility of the reactants to substitution—in this particular case, Cr^{2+} (aq) is extremely rapid in its substitution reactions and step 1 will be fast.

Step 3 (decomposition of the complex) need not necessarily, involve a ligand transfer: $[IrCl_6]^{2-}$ when reduced by an inner sphere mechanism, retains its coordination shell intact. In the example given above, the chloride is retained by the Cr^{3+} ions as Cr(III) complexes undergo substitution reactions slowly. Co^{2+} forms very labile complexes, and will readily give up the chloride ligand and lose its coordinated ammonia ligands.

For the initial step to be favourable, it is clear that at

least one of the reactants must be substitutionally labile, and that a suitable bridging ligand must be available. Thus $[V(OH_2)_6]^{2+}$, which undergoes substitution more slowly than $[Cr(OH_2)_6]^{2+}$, will often react by an outer sphere mechanism when Cr^{2+} aq reduces via an inner sphere process.

The role of the bridging ligand

Clearly the formation of a bridged complex has the advantage of bringing the reactive centres close together, but it also appears to favour the electron transfer in other ways. For a simple bridging ligand such as Cl^-, the bridging appears to increase the interaction between the donor and acceptor orbitals, and to reduce the energy barrier through which the electron has to tunnel, A simple theory is discussed by Halpern and Orgel. The case of remote attack is somewhat better understood. It has been found that only ligands with delocalised π systems joining the two metal binding sites may act as bridging species, implying some sort of conduction through the ligand.

In the case of ligands which are relatively easily reduced, it has been shown that the reaction 2 appears to take place in two steps, an initial reduction of the ligand by the reductant, followed by a reduction of the oxidant by the reduced ligand species. The symmetries of the orbitals involved also play an important part in determining the rate. The rates of reduction of some pentammino (4-carboxyamino-pyridyl) complexes by Cr^{2+}aq are:

Metal	*Configuration*	*Rate constant k sec⁻¹ M⁻¹ at 25°C*
Cr(III)	t_{2g}^{3}	1.8
Co(III)	t_{2g}^{6}	17.4
Ru(III)	t_{2g}^{5}	3.8×10^5

All these reactions will pass through a bridged intermediate similar to (4). The similarity of the rates for Co(III) and Cr(III) when the total free energies of reaction differ by 60kJ/mole may be explained by assuming the initial electron transfer step involves reduction of the bridging ligand.

The much greater speed for Ru(III) reduction is explained by the fact that the reducing electron enters a t_{2g} orbital and not an e_g*, and the t_{2g} may overlap and mix with the ligand π system. The reduction of Ru(III) by the reduced ligand is thus greatly accelerated. It seems reasonable to imagine that t_{2g}—t_{2g} electron transfer will be most successfully mediated by a p system on the ligand, and that e_g*—e_g* transfer will favour passing through a s system.

We might regard the single atom bridging species such as Cl^- as examples of a s system. There is often a considerable trans- effect for transfers to e_g* orbitals, strong ligands trans- to the bridging group slowing down the reaction, presumably as a result of the greater reorganisation energy when the trans- ligand interacts strongly with the e_g* orbital.

Factors favouring inner sphere or outer sphere paths

Very frequently the inner sphere and outer sphere paths have roughly equal probability, and it is not uncommon for both to contribute to the reaction. Species with no bridging ligands, with low reorganisation energies, or with very slow substitution rates will tend to favour outer sphere mechanisms, while substitutionally labile species will be more attracted to inner sphere paths. An inner sphere mechanism may offer a means of reducing reorganisation or Franck-Condon barrier too large for an outer sphere mechanism to be followed readily.